DANCING IS
THE BEST MEDICINE

**JULIA F. CHRISTENSEN
& DONG-SEON CHANG**

Translation by
KATHARINA ROUT

DANCING
IS THE
BEST
MEDICINE

The Science
of How Moving
to a Beat Is Good for Body,
Brain, and Soul

GREYSTONE BOOKS
Vancouver/Berkeley/London

First published in English by Greystone Books in 2021
Originally published in German as *Tanzen ist die beste Medizin*,
copyright © 2018 by Rowohlt Verlag GmbH
English translation copyright © 2021 by Katharina Rout
Illustrations by Julia F. Christensen

22 23 24 25 26   6 5 4 3 2

All rights reserved. No part of this book may be reproduced, stored in a retrieval system or transmitted, in any form or by any means, without the prior written consent of the publisher or a license from The Canadian Copyright Licensing Agency (Access Copyright). For a copyright license, visit accesscopyright.ca or call toll free to 1-800-893-5777.

Events, individuals, and places mentioned in this book are based on real events, individuals, and places. They have, however, been altered to fit the narrative.

Greystone Books Ltd.
greystonebooks.com

Cataloguing data available from Library and Archives Canada
ISBN 978-1-77164-634-5 (pbk)
ISBN 978-1-77164-635-2 (epub)

Editing for the English-language edition by Linda Pruessen
Proofreading by Jennifer Stewart
Cover design by Jessica Sullivan and Fiona Siu
Text design by Fiona Siu
Cover illustrations by Macrovector / Shutterstock

Greystone Books gratefully acknowledges the Musqueam, Squamish, and Tsleil-Waututh peoples on whose land our Vancouver head office is located.

Greystone Books thanks the Canada Council for the Arts, the British Columbia Arts Council, the Province of British Columbia through the Book Publishing Tax Credit, and the Government of Canada for supporting our publishing activities.

Canadä

CONTENTS

A Note From the Authors   9
Prologue   12

CHAPTER 1: SOLO DANCE–RHYTHM I CANNOT RESIST   17

> We need this. This is not a luxury—it is a necessity.
> **SIR SIMON RATTLE**

The Success Story of Dance   18
Magic Rhythm   23
Born to Dance   31
End of the Line: Prom   34
The Pony Farm Dance   43

CHAPTER 2: PARTNER DANCE–DO YOU SPEAK DANCE?   51

> Dance is Esperanto with your whole body.
> **FRED ASTAIRE**

Something in the Way She Moves   52
Mirror Dancing   55
May I Have This Dance?   65
The Search for a Partner   70
To Lead and Be Led   76

CHAPTER 3: GROUP DANCE— THE SOCIAL BENEFITS OF DANCING 82

Dance is the only art of which we ourselves are the stuff of which it is made.
TED SHAWN

Dancing for the Feeling of Community 83
Moving With the Times 89
Team Spirit 94

CHAPTER 4: DANCING FOR THE BODY— DANCE AS AN ELIXIR OF LIFE 97

[When I dance], I sort of disappear, I feel a change in my whole body…there's fire in my body…I'm just there… I'm flying like a bird, like electricity.
FROM THE FILM *BILLY ELLIOT*

I Like to Move It, Move It! 98
A Very Special Nerve 101
Music in Our Head 105
Body Perception 110
Dancing Makes Us Smart 114

CHAPTER 5:
PRESCRIBE DANCE, NOT DRUGS 122

*Wherever the dancer steps, a fountain
of life will spring from the dust.*
RUMI

The Effects of Dance on
Our Heart and Immune System 123
Back and Joints 128
Weight Loss 133

CHAPTER 6:
DANCING AS THERAPY 139

If life brings you to your knees, do the limbo dance.
ANONYMOUS

Dancing Our Emotions 140
Dancing Away Your Inner Couch Potato 147
Stress: Bitten by the Spider 153
Against Anxiety 160
Depressed? Dance It Out! 164
If One of Our Senses Is Missing 171

CHAPTER 7: YOUNG AND OLD–
DANCING AT ANY AGE 177

On with the dance! Let joy be unconfined.
LORD BYRON

Dancing to Protect Against Memory Loss: Dementia 182
Parkinson's Disease 186
Dancing for Children 189

CHAPTER 8: DANCE DOES EVEN MORE— LET'S LAUGH, CRY, AND DANCE 194

Sometimes you get lucky and find a soul that dances to the same beat as you do.
ANONYMOUS

This Is Also About Sex 195
Nice to Meet You 201
Enjoying Dance From a Distance 206
Integration and "Accep-dance" 219
Let's Dance 227

CHAPTER 9: DANCE TEST: WHICH STYLE FITS ME? 230

You got to dance like nobody's watchin'.
SUSANNA CLARK

It Has to Fit 230
Checklist: What Do You Need for a Dance Class? 233
May I? Dance! 234

Acknowledgments 254
Sources 258
Index of Health Concerns and Dances 299

A NOTE FROM THE AUTHORS

DEAR READER,

When we wrote the German version of this book in 2018, our aim was to share our excitement about our favorite hobby—dancing—and to get everyone dancing along with us. "We'll tell everyone about the amazing scientific discoveries around dancing's health effects," we thought, "and that will make the whole world get out of their chairs and onto the dance floor!" We couldn't wait!

But when the time came for the English version of the book to be published, the world was in the midst of a pandemic, and dancing was forbidden.

This, too, shall pass, goes a Middle Eastern saying. The words come from an inscription on a mystical ring once gifted to a king to boost his morale. And they are so true, even for bans on dancing.

Dance has known many adversities throughout recorded history… over and over again, kings and political leaders have banned dance for social or political reasons, or both; dictators have banned assemblies out of fear of the crowd, punishing dance communities in the process; and so-called spiritual leaders have banned dance because it supposedly corrupts our mind. Dance has been accused of instigating social unrest and disease; dancers have been beaten with sticks and whips and called names like "degenerate." Dance has been forbidden, considered a disgrace.

Sometimes, it's our own life choices that make it so difficult to dance on a regular basis. Responsibilities (a busy family life that leaves no time for a hobby), geographical challenges (the nearest dance school is on the other side of the city), and even shyness (dancing in public is so embarrassing!) can get in the way...

What does all of this mean? There's always a reason we "can't" dance: sometimes it's personal, sometimes it's political, and sometimes it's because of a pandemic. However, none of this should stop you from taking dance classes, or keep you from grooving around from time to time.

History and those words on the king's ring teach us that obstacles come and go. With this in mind, it's best to keep our dance shoes and knees ready for the next dance party, because all of this, too, shall pass.

And, for now—and in preparation for all the adversities to come—it's our pleasure to share what has been the antidote to all of the past obstacles to dance:

Dance. *Just keep dancing.*

At home, alone, in a basement, in secret, in a lockdown?

Yes.

Life is constant improvisation, and so is dance and dance practice. Dancing helps you broaden your movement repertoire—and your mind. It makes your back and your neurons more flexible, and enhances your problem-solving skills. You may think twice about trying to solve the problem of learning a basic samba dance step in your living room while navigating the dog's tail, the constant online meetings, and the challenge

A Note From the Authors

of buying groceries when a stay-at-home order is in place. But your brain's ability for neuroplasticity—that is, the ability to form new connections—will be firing up the engines to make you mentally fit for the future. Just keep at it. Keep dancing. At home, in the basement, at the bus stop, and especially in your mind.

One good thing brought about by the COVID-19 pandemic is that online dance possibilities increased exponentially. What would normally take years to develop, got up to speed, following the beat of the moment, in a few months. Suddenly, Julia was able to participate in Argentine tango classes again, with her Argentine tango teacher, Raquel Greenberg, in London. And Dong-Seon, in South Korea, was back on the dance floor with Dax and Sarah's Rhythm Juice and with Jo and Kevin's iLindy from California. It makes us so happy! And that's a good thing for our health. As science shows (and as you'll see in the pages to come), positive feelings activate our body's relaxation response and give our immune system a boost. Want to give online classes a try? Check out the section on online dance in chapter 9 for some tips to get you started.

In the meantime, know that we're just as enthusiastic about our hobby as we've always been—and just as excited to share the news about its many physical and social benefits with you. We think you'll be surprised by what you learn here, too, and that you'll soon be eager to join us on the dance floor—wherever that may be!

Remember dance?
That move, that feel, that beat?
Remember to remember,
to dance.

—DR. JULIA F. CHRISTENSEN AND DR. DONG-SEON CHANG

FRANKFURT AM MAIN, GERMANY / SEOUL, SOUTH KOREA, JANUARY 2021

PROLOGUE

ARE YOU UP for a dance? We dearly hope so... We? Dr. Julia F. Christensen and Dr. Dong-Seon Chang—neuroscientists by day, dancers by night, and the authors of this book. Nice to meet you!

I'm Julia, from Denmark. My life's dream was to become a dancer. I started ballet when I was very little, and like many girls, I hoped that one day I'd be dancing on the world's biggest and brightest stages. My whole life was about dance. I practiced every afternoon after school and started professional training after graduating from high school. But then life took an unexpected twist. I injured my back in an accident, and it was so bad that I had to give up my dream. Changing plans was hard—really hard—but I did it. I turned to another thing I'm very passionate about: human behavior. I studied psychology and neuroscience in France and Spain, and eventually received my PhD from the University of the Balearic Islands on Majorca. For some time after the accident, I didn't want to have anything to do with dance. The memories were simply too painful. But then, in the early days of my research career, I found myself studying a very particular type of human behavior: dance! And all of a sudden I was back in the world I'd loved so much for so long—but looking at it through an entirely different lens. Dance became my area of research. I learned that although dance is often considered a simple courtship ritual for the attraction of mates, it's actually so much more. Dance is an antidote to stress, a way to combat negative emotions, an elixir for the body, mind, and brain. Amazed by what science was teaching me, I slowly ventured back into the dance world to savor the effects of recreational dance firsthand...

Today I live in London. By day, I'm a neuroscientist, researching and publishing on dance and the brain. But when night falls, I head out into the city to dance the Argentine tango, enjoying the health and social benefits of this delightfully strange and uniquely human behavior. Against all odds, I've ended up back on the dance floor, and I couldn't be happier.

· · · · · · · · · · · · ·

I'm Dong-Seon. I was born in Heidelberg, Germany, to Korean parents. After graduating from high school, I studied biology in Konstanz, Germany, and cognitive science in New Jersey, in the United States. Movement was at the core of my research for my PhD in Tübingen, Germany. I examined how the brain perceives human movement and human interactions. At some point, I discovered one of the most amazing types of human movement interaction: dance! Unfortunately, I then entered a difficult period. Overly ambitious and in search of something to give real meaning to my life, I struggled with serious depression. Luckily, I had a good doctor—and destiny gave me another present: swing dance! Together, my doc and swing dance allowed me to discover dance as a medicine. They gave me back my joie de vivre and my energy. Ever since I was a kid, music had played a key part in my life. My instrument is the piano—me and my piano; that's when I really unwind. Surfacing after this depressive episode, I realized that dance created a bridge between the two things that mean the most to me: music and movement.

Today I develop innovative human-machine interaction technologies at an institute for Research of the Future; as a science slammer, I present the latest findings in neuroscience on stage; and occasionally I perform as an amateur musician and DJ. I'm also a father, and the husband of a wonderful woman who, unfortunately, doesn't want to dance with me. I'm crossing all my fingers that this book will change her mind!

• • • • • • • • • • • • • • •

ENOUGH ABOUT US. Let's put dance center stage.

Did you know that in ancient Greece, Apollo was the god of dance and music, *and* of healing? And did you know that this combination of dance and healing has an important place in many mythologies? Wherever we look, gods and goddesses of dance are linked to healing and health. The Egyptian goddess Bast (or Bastet), the Semitic god Baal, and the Hindu god Shiva—all are in charge of dance, health, and healing. Without the benefit of any neuroscience at all, prehistoric humans understood the important link between dance and health. And dance has played a central role in the healing rituals of many cultures ever since: in the Brazilian rainforest, in the Kalahari Desert of Botswana, in East Asian Buddhism and shamanism, in classical Indian dances, and in the rituals of ancient Greece, with the patron god Apollo.

Much to our surprise, we discovered that Apollo is also the name of a hotel on the Greek island of Aegina. Every year, a huge crowd of neuroscientists from all over the world gathers here for a conference about "the human social brain." This is where we—Julia and Dong-Seon—first encountered each other. A wink of destiny? Hotel *Apollo*? Really? We met at the hotel bar one night, watching our colleagues groove away on the dance floor. Not surprisingly, with this spectacle in front of us, our very first conversation was about dance and the brain. What happens in our brain when we dance, and why does it make us so happy? Can dance increase our health and well-being? Make us smarter? Help us make new friends? Our conversation went on for hours. When we'd had enough of talking, we plunged into the crowd and danced the night away. After what was likely too little sleep, we joined the conference lectures the next morning. But that casual chat at the hotel bar turned out to be the kickoff for this book: for eight consecutive days and nights, we alternated between scientific debates about the brain and sessions on the dance floor. Slowly, as conversation and dance converged into a single train of thought, the chapters of this book emerged. Have a good read. And when you're done—or maybe while you're still at it—go out and *dance*!

1

SOLO DANCE– RHYTHM I CANNOT RESIST

••••••••••

We need this. This is not a luxury—it is a necessity.
SIR SIMON RATTLE

••••••••••

Rousing summer hits come from the loudspeakers. We are standing at the outdoor bar of our conference hotel catching our breath. The first day of a scientific conference is always exciting, especially if you happen to be in a foreign country. The trip, the hotel, colleagues from all over the world–it's a lot to take in. And then everyone plunges right in to scholarly lectures, panel discussions, and the exchange of information. How often does one get the opportunity to talk with so many colleagues who specialize in the same topics? These conferences always begin with a series of introductions that the organizers like to keep informal and friendly. *This particular day started with a little contest: "Whose topic gets the most Google hits?" "Social Neuroscience"—the overarching topic of the conference—got a measly 8,730,000 hits. "Empathy" did better, with more than 50 million hits. But Julia's research topic outperformed them all by a huge margin: the word "dance" returned 1,820,000,000 hits.*

THE SUCCESS STORY OF DANCE

In May 1927 the American pilot Charles Lindbergh made a nonstop solo flight from New York to Paris. Above the stormy Atlantic Ocean, Lindbergh and his single-engine *Spirit of St. Louis* were dancing in the clouds. Down below, in the clubs of Harlem, people were dancing too—and their dances were no less stormy! The Charleston, jazz dance, and the breakaway were the favorites of the day, but a new style was also taking over the floor. It was a combination of everything that had come before it, and its step sequence was the result of chance: in one of the many dance socials taking place all around the city, partners Mattie Purnell and George Snowden were dancing the breakaway, a swing dance that had originated in the African American community. As they let themselves be carried away by the music, a new step sequence emerged. Again, and then over and over again—as if the rhythm had taken control of their legs. Soon, the other dancers on the floor had formed a circle around them. Unable to take their eyes off the couple, they watched curiously and cheered them on. In the early hours of the morning, when a sweaty George Snowden was asked what he called this new dance, his eyes caught a glimpse of a newspaper account of Lindbergh's transatlantic flight. *Lindy Hops the Atlantic*, the headline read. Casually, and with a big grin, Snowden replied, "Lindy Hop." Lindbergh's dance in the clouds had turned into a dance on the floor. A new hero and a new style of dancing had been born: Lindbergh and the Lindy Hop. (There are many different stories about how exactly the Lindy Hop came to be, but in each of them, Mattie Purnell, George Snowden, Frankie Manning, and Norma Miller play the central roles.)

What made this new dance particularly special was the option of dancing solo or as part of a couple. Alone with the elements and oneself—that's a powerful experience. And for many, crossing the dance floor is just as nerve-racking as crossing the Atlantic must have been for Lindbergh. Dancing feels exciting, and it always leads us into uncharted territory, depending on the partner, the dance hall, or the music that happens to be playing. Dancing can be a ritual, a sport, an art form, a profession, a passion, or therapy. But above all else, dancing is an expression of feelings. Children dance to "Soft Kitty" while Grandma sways to folk songs in her seniors' residence. A long-haired metal fan rocks with the beat at an open-air festival while a ballet dancer twirls on a city stage in her pink tutu. Everybody does it. *Everybody* dances.

Dancing has sunk its roots deep into all of us. As soon as we hear a familiar rhythm, we want to move, even if it's just to bob our head. The desire to dance, it would seem, is rooted in our brain. As neuroscientists, we'd go so far as to say that the brain wants to dance! And dancing is probably as old as humanity itself. A quick look through human history reveals how important dancing has been for the development of our species, even though at first glance rhythmic movement appears pointless from an evolutionary perspective. Early humans inhabited a harsh world of deprivation and danger. It took a great deal of time and energy to provide sufficient food and ward off enemies, be they human or animal. Boredom did not trouble our long-ago ancestors, and when it got dark, they were in all likelihood exhausted from the work of the day. So why did they "waste" their hard-earned energy reserves on dance? Because make no mistake: dance they did!

   Prehistoric humans danced in times of mourning and in times of joy; they danced to conjure rain, to appease the gods, and to stoke their rage toward their enemies. We can see how important dancing was to our ancestors when we look at their rock paintings—Stone Age graffiti, if you like. Four topics are regularly represented: animal and hunting scenes, family and property depictions, sexual behaviors—and dancing. And dancing was most likely practiced long before humans were capable of keeping records of any kind. In an interview with *Der Spiegel*, music cognition researcher Gunter Kreutz of the University of Oldenburg described dance as a by-product of the erect posture that human beings adopted and that contributed tremendously to the development of our cognitive abilities: "Perhaps humankind developed as much as it has thanks only to dance."

   When we look at evolutionary history, we find clear indications that it was only our own species, humans, that at some point started to make music. Of course, our ancestors didn't relax in their caves to listen to Mozart's *Requiem* on their earbuds; for them, music always meant being active. It involved movement and an engagement with what was on their minds and in their hearts. Since prehistoric times, humans have been moving to rhythm or music to express with their bodies aspects of their lives and their everyday practices.

Unfortunately, no choreographies from the early days of humankind have come down to us; dance does not fossilize or leave behind musical instruments or other cultural artifacts that we can admire in museums. Dance disappears the moment it is danced—it remains only in the memory of the dancers and their audiences. Even so, the ritual dances of modern Indigenous peoples allow us a glimpse into the importance of dance for our ancestors.

Cultural and art historian Aby Warburg described the snake dances of the Pueblo Indians in the early twentieth century, intended to conjure rain. Tribesmen took the heads of live poisonous snakes into their mouths and held the reptiles between their teeth while they danced. The earth needs water, the thinking went, and because snakes are closer to the earth and hence to the powers of nature, they are better at communicating with the clouds than humans are. With the snakes in their mouths, the dancers were able to transfer the tribe's message to the clouds: *we people need rain.*

Of course, we now know that these dances had absolutely no effect on the rain. But our ancestors—who existed without the benefit of scientific explanations for the phenomena of their daily lives—had no idea why it sometimes rained and other times did not. Why were there seasons? Would summer return? What if it didn't? Would morning always follow night? And why were there sometimes horribly loud thunderstorms? Were they messages from furious gods?

For early Indigenous peoples, dance was a means of exerting control over the mysteries of their everyday lives. If we consider the uncertainties and insecurities they faced, we can understand why our ancestors developed strategies to give themselves a sense of control, and the feeling that they were able to do something to secure their tribe's survival. Independently from each other, many Indigenous peoples created dances to honor or appease their gods, to prepare for the hunt or for battle, and to engage with the weather and influence the harvest. Dance, for them, was an important ritual.

Evidence that dancing bodies have always captivated our imagination can be found in prehistoric rock paintings, in the countless Google hits that popped up when Julia searched the term "dance," and in the high ratings of television programs such as *World of Dance* and *Dancing With the Stars*. People all over the world—regardless of culture or background—move to the rhythm of music, and they do so from early childhood to advanced old age because dance is movement like no other. Dancing ranges from simple rocking movements to highly complex art forms such as classical ballet.

Dance is unique. Most movements have a purpose. In our daily lives we move in order to get from point A to point B, to complete a task, to communicate, or to have a particular impact on our environment. We nod or shake our head, close a door, wave, or point at something in order to make ourselves understood. Movement in sports, too, is goal-oriented: we overcome a particular distance in a certain amount of time, or we kick a ball into a net. Mostly, movement in sports is about competition—comparing ourselves to others and trying to do better than they do.

Dance, though, is different. Dance movements arise from within and are not concerned with what is happening around us; they do not care about any effect they might have on the world. Purposeless, they are an end in themselves, an outward expression of an inner state. The outstanding health effects are, for the most part, a completely unintended side-effect. We don't need to think about doing anything in a particular way;

when we let the music take control, the health effects just happen. Dancing is, by definition, a spontaneous movement to a rhythm. That impulse to follow the rhythm originates within us, and every person will respond in their own way. In order to throw a ball, we practice movements that are similar all over the world; few variants exist. But in order to dance to a particular piece of music, we can move in as many different ways as there are people in the world. Some skip and pulse to a strong beat while others lean toward flowing movements; some move mostly their arms, others their legs. Some close their eyes, others don't. This freedom of expression is unique. Wolfram Fleischhauer describes this beautifully in his book *Fatal Tango*. A ballet dancer is searching for a tango dancer in Buenos Aires, a man with whom she had fallen in love in Germany and who had then left the country after their brief, passionate encounter. To find him, she watches dancers and approaches those whose movements are similar to his; she thinks they might have learned dancing from him. By studying his dance movements, she hopes to also discover his identity—a quest that sets off a suspense-filled, intriguing story.

No matter how hard it is to define the word *dance*, we know it when we see it. Dance may be a ritual with strict rules, a precise art form, or a sweat-inducing sport, but above all it is this: a feeling primarily related to our inner lives.

MAGIC RHYTHM

The first day is drawing to a close. We've already listened to many interesting papers and have enjoyed meeting so many people who are exploring the same or similar questions: What makes us human? What is going on in our heads that makes humans so unique? What happens in that jelly-like organ that weighs only about three pounds, works faster than most computers, and is likely to remain a mystery no matter how much we research it?

Slowly, darkness is falling. We order another round of drinks and discuss dancing while others simply dance. More and more people have hit the dance floor, and the DJ gives his all to motivate the last holdouts. The music is so loud now that it's hard to talk. The beat grips us too; we tap our feet, and our heads bob in time.

"Do you think there are people who can completely ignore rhythm?"

"No. Our brain wants to dance. We have no choice."

Many researchers of cognition and evolution agree that classifying art as music, dance, or singing is a very modern invention, and that for most of human history the arts were at the center of social life and not separate from each other. We can see this unity reflected in languages that to this day have only a single word denoting both music and dance. The Greek origin of our word "music"—*mousike*—referred simultaneously to music and dance, while the Greek for "rhythm" referred to movement and was not limited to the musical aspect. This makes a lot of sense, for research has shown that music and dance—or listening and moving—are firmly connected in our brains. Dividing them into different categories is a violation introduced more recently by a society shaped by the dominance of reason we inherited from the Enlightenment.

When I started my work as a researcher in London, I went to the imposing and world-famous Royal Albert Hall. Daniel Barenboim, the well-known conductor and pianist, was at the piano. It was breathtaking. The music filled me; it flowed through my body as I sat in this venerable concert hall and tried to stop myself from swaying along. A few times I couldn't hold back, but my neighbors' stern looks reminded me that "you don't do that." I was astounded at the discipline of this audience that was as quiet as a mouse and only clapped, coughed, or cleared their throats during the pauses between the pieces.

It was an entirely different experience during another visit, when the Persian artist Adib Rostami took to the same stage. This time, the entire audience let itself get carried away; no one sat quietly in their seats. They were engrossed in the music. They laughed and even cried; some got up and danced in the aisles.

In both cases, my brain recognized the rhythm without having to think. But what we do with our bodies when we hear a rhythm depends on our personal history—on what we learned while growing up and on our cultural background. In a Western culture, our social environment teaches us that music and dance are separate. We learn to sit still through a concert or ballet performance (and to find that entirely normal!) instead of doing what our brains would otherwise do automatically; namely, join in.

When we listen to music, we often cannot help but move. You, too, have probably tapped your foot in time with music without really meaning to. This happens because the nerve cells in our brain responsible for listening and controlling movement are linked. Sounds from our environment enter our ears and travel as nerve impulses for movement straight into our legs. Studies have shown that strong rhythms can sometimes be measured as minimal muscle contractions in our arms and legs. We are born with these connections, which are then reinforced by the behaviors of the people around us in our first years of life: to put babies to sleep, we rock them and sing lullabies, and we clap and stamp our feet when we sing with children. Our brains translate sounds into impulses for movement. The sound waves of our favorite music take charge and command our muscles to move.

Yet the urge to move to the sound of music is different for each individual. Some enjoy a waltz, some prefer AC/DC's "Hell's Bells," and others feel the need to move their feet to Justin Bieber's music. Researchers have discovered that every one of us ultimately prefers a particular individual groove; the more we like a piece of music and the more clearly we discern its rhythm, the more intensely we sense our individual groove.

For dancing, we don't necessarily require a melody, but we always need a rhythm. And it's not enough to have just any kind of sound. If it's too shrill or loud, "detectors" in our brain signal danger (in the same way they do when we hear a siren, for example). We need pleasing sounds and a sequence of tones and pauses: a rhythm. Rhythm gives a piece of music a temporal order because it repeats at regular intervals. We recognize it even when the melody seizes on something entirely new. In music, the interval during which the same rhythmic figure occurs is called time, and whether we know it or not, we've all sensed it at some point—maybe at a concert, a wedding, or just while driving along listening to the car radio. You may even have counted along in your head—one-two-three, one-two-three… for a waltz, for example, which is three-four time, meaning there are three beats per bar of music. Many other times exist, but in the West, three-four time and four-four time are the most common. And within these times the rhythm unfolds. Often it is played by the drummer and additionally emphasized by the percussionist and the bass player.

But rhythm is not the same as time: rock 'n' roll, for example, has an entirely different rhythm than rumba, even though both are played in four-four time. Dance and rhythm are inseparable, with rhythm forming the link between music and dance. Sir Simon Rattle put it this way in his famous project *Rhythm Is It!*: "Rhythm is not a luxury but a need, like the air we breathe." You may think you don't have a feel for rhythm, but trust us when we tell you that's impossible. Our whole life follows a rhythm. Rhythm is entirely natural. We find it in minutes and hours, in day and night, in the moon and the tides of the ocean. Rhythmic processes happen in our bodies all the time. Our heart beats in a particular rhythm; we breathe in and out; and we sleep and are awake according to a specific temporal pattern. Even our brain has a rhythm. Neuroscientist György Buzsáki of New York University has found that the nerve cells in our brains create rhythmic vibrations. In order to master complex tasks such as cognition, emotions, and language, our neurons must dance, in a manner of speaking, to the same rhythm.

Much of what we do happens according to rhythms of which we are entirely unaware, and that's a good thing. Imagine if our brain were incapable of recognizing regularities in our environment. Each event would be a first and would need to be thoroughly assessed, since everything could potentially be dangerous. Everything would have to be monitored at all times. It would be hugely exhausting. This is why our brain is able from birth to recognize regularities as rhythm. The sun comes up; the sun goes down—day and night, alternating roughly every twelve hours. We know this, and so we don't need to be frightened when it grows dark. As soon as we have seen or heard a particular rhythm a few times, our brain recognizes it, can anticipate it, and assesses it as not dangerous.

The beat we know best is the rhythm of our feet when we walk: one-two, one-two, one-two. Spend a few minutes listening to the music played by a marching band at a parade and it's easy to hear how it draws on the natural beat of our gait.

When people in the eighteenth century heard a waltz for the first time, three-four time was a novelty for their brains: *one*-two-three, *one*-two-three... It was new and exciting. But once they'd heard it a few times, their brains memorized it, just the way they learned to recognize the rhythm of the sun rising and setting, or of their own feet on the ground.

When you hear a new rhythm for the first time, give your brain a chance to take it in. Listen for a moment, and perhaps even mimic it with your hands. Follow the rhythm of a waltz, for example, by gesturing waves or by clapping along. As soon as your hands can do it, your legs will find it easier, too.

In a 2009 study, psychologist István Winkler of the Hungarian Academy of Sciences in Budapest arranged for fourteen newborn babies to listen via headphones to a composition for drums and bass while they were sleeping. In the rhythm of this particular piece, the first beat carried the stress. Winkler discovered that any time that first beat was left out, the babies' brain waves changed. Newborns already recognize rhythmic patterns and react to changes in the music.

Many scientific findings demonstrate that our brains love predictable rhythms. And yet... most hits and catchy tunes will notably break from their predictable pattern at some point. If rhythm becomes too monotonous, we get bored and stop listening.

American neuroscientist and professor of music David Huron wrote a book on this topic. Its title, *Sweet Anticipation*, hints at the fact that our species loves recurring elements that create an anticipatory mindset. If a piece of music deviates from what we expect—for example, by shifting the stress in so-called syncopation—our brains are surprised. We can see clearly, and even measure, changes that occur in our brain waves at that moment. The changes are caused by the neurotransmitter dopamine.

This reaction can easily be explained by evolution. As human beings developed into ever more complex creatures, every deviation from predictability deserved our ancestors' special attention. Different was, after all, potentially dangerous. To this day, dopamine motivates us to notice changes in any given rhythm. Too much unpredictability, however, is not every brain's thing—which is why not everyone loves complex jazz. It's all about the right balance. Canadian psychologist and professor of philosophy David Berlyne researched this balance and found that most people prefer a medium level of complexity and deviation: neither too much nor too little.

The achievement of this balance is exactly the art that characterizes a good DJ. Depending on the evening, the mood, or the audience, a DJ chooses different music to lure people to the dance floor. Unfortunately, there is no magic formula that works for everyone. When I moved to Berlin, I met a number of well-known DJs and was allowed to watch for long stretches as they worked and experimented. One of my most important observations was that songs that bring people to the dance floor have a few things in common: they have to be widely known, and their beat has to be identifiable enough for people to bob along. First, the DJ has to get more than three or four people to move in time. But then "funky beats" are needed, beats that are not completely regular but come with a small delay, or syncopation. At this point, a good atmosphere often turns into enthusiasm, and the dance floor fills.

In 2016, Dr. Maria Witek and her colleagues at Aarhus University in Denmark wanted to find out more about this phenomenon. They designed fifty short percussion breaks in four-four time with varying numbers of syncopations in order to create rhythmic tension. The research participants perceived these breaks as variously complex and variously likeable. In addition, Witek used the motion sensor of a Wii console to measure the participants' upper-body movements during those breaks. What do you think she found? Were the participants able to stop grooving? Of course not. Whenever a break was played that had the right mixture of anticipation and surprise, the dancers started grooving. Some moved minimally while others exhibited more expansive movements, but none was able to remain still!

We can see the magic of syncopation in a YouTube video we came across in our research. A dashboard camera records three Italians in a car. With expressive Italian gestures they sound off about the smash summer hit "Despacito": they cannot stand it! They hate that the song is played everywhere and all the time, and they hate how everyone starts dancing the moment the song is played. The song is a disaster! Then the car radio plays the next song: "Despacito"! The Italians keep right on griping about it, but at the same time, they start, on and off, to sing along, bob their heads, and groove. They simply cannot resist the music's effect, even though they absolutely hate the song! Every year there's a summer hit that, at some point, we can't bear to hear anymore. But when it gets played on the radio for the umpteenth time, we will still catch ourselves starting to move.

BORN TO DANCE

The first bars of "Uptown Funk" by Mark Ronson and Bruno Mars can be heard from the hotel bar. A funky beat if there ever was one! The dance floor is packed. The DJ knows what he is doing; he's creating the perfect anticipatory mood in his audience of vacationers and scientists. Even the professors who always look so serious are dancing.

"Is there an audience you can always get to dance?" I ask Dong.

He doesn't need to think long: "My son!"

....

As a baby, Theo was always ready for action exactly when we were utterly exhausted. He would scream and cry, fidget and whine, and demand our complete attention. Only one thing would do the trick: swing dancing! The magic combo for Theo was being carried along as we did the Lindy Hop to a fitting song such as "Dream a Little Dream of Me" or "On the Sunny Side of the Street." He would stop crying, look pleasantly surprised, quiet down, breathe more calmly, and allow himself to be danced into sleep. Fortunately, Theo was not picky; jive, jitterbug, or rock 'n' roll all worked well—as long as it had the step-step-rock step. I have no idea how many nights I danced with him through our apartment to make him fall asleep.

When he was a bit older, he fell in love with the soundtrack from *Sing*, an animated children's movie in which animals perform at a singing competition. His favorite was "Bamboleo," sung by a herd of pigs, but he also loved "Don't Stop Me Now" by Queen. The latter became his motto for life, and as soon as he was able to stand, he skipped, clapped, and headbanged as much as his little body would let him whenever this song played. Getting him to dance is super easy!

Most parents probably have similar stories about their children. But what appears to be so natural at first glance is actually somewhat unusual in the animal kingdom. Hardly a species besides humans seems capable of responding spontaneously with rhythmic movement to sensory stimuli from its environment. The phenomenon is the same all over the world. Embryos in the womb respond to rhythms with movement, and this impulse to move rhythmically continues through childhood, whether two-year-old Leon bangs his head to Metallica's booming bassists or ten-year old Charlotte, lost in movement, dances in her living room to the sounds of Chopin. And it works equally well for Dong's son with his Asian roots as for Julia with her Nordic genes.

My father has always loved to listen to music. And he likes his music loud. So loud, in fact, that my mother bought me ear protectors shaped like a couple of pink rabbits. I was about two when we went to a hi-fi fair. The basses were booming, and the visitors' feet and heads were bobbing to the beat. I was sitting in my stroller with my rabbits on my ears while my father was completely in his element. To my mother's everlasting shock, I made the pink rabbits dance. A particularly stylish entertainment center had claimed all of my parents' attention. When they eventually turned back to me, I had adopted a dance position. Kneeling in my stroller and holding my rabbits in my hands (rather than wearing them over my ears), I was bobbing to the sound of the music: my head, my body, the rabbits, and the stroller itself were all moving up and down to the beat. The stroller's springs were picking up the rhythm and gave the whole thing just the right oomph. A small audience had gathered in a circle around me. "Look at that kid! She's dancing!" they exclaimed. No one had taught me or asked me to–the boom from the boxes had been enough to make me dance with my rabbits. To the people around me, what I was doing was completely clear. They didn't call it hopping, seesawing, or jumping, but dancing. There was no doubt in their minds.

Music had grabbed hold of little Julia. She couldn't help it. And science confirms it: several studies show that babies and toddlers cannot suppress the urge to dance. We can't stop them, and shouldn't try to. If they want to dance in a stroller that's a bit rickety, we'd better pick them up, lift them out, and dance with them... At birth, our brain lacks many of the important capabilities we will develop through elaborate learning processes over the course of years. But the impulse to respond to rhythm with dance is already fully developed in newborns, as researchers at York University in the United Kingdom discovered in collaboration with colleagues from the University of Jyväskylä in Finland. Of course, the motor skills of newborns don't allow them to impress us with complex moves on the dance floor. We therefore call it "dancing" if babies spontaneously and without prior demonstration by others synchronize their movements to a rhythm. In a study involving 120 children between the ages of five months and twenty-four months, researchers observed that the little ones started to move as soon as they heard music or drumbeats, and that their movements grew stronger the more clearly identifiable the rhythms. The babies bobbed their arms and legs, and so often did so in time that it could not be explained by chance. However, they only danced to rhythmic sounds. When babies were exposed to random sounds or language, they did not respond with movement.

The sense of movement is thus innate. This is astounding, given that human babies, in contrast to many animal babies, are utterly helpless at birth—and yet they know right away how to dance, or move along to a beat. By the way, researchers also found that babies smile when they dance. From the day we are born, movement to music appears to give us joy.

END OF THE LINE: PROM

We are watching a colleague who, with glass of wine in hand, is leaning against the wall at the edge of the dance floor and watching the dancers. Earlier today, he gave a fascinating lecture about coordination and communication and about the importance of participating in shared activities for social bonding.

"Do you see that, Julia? Crazy that he of all people is standing so alone and forlorn on the sidelines. He struck me as open and outgoing."

"I don't think he's introverted. I do think he knows how

Solo Dance—Rhythm I Cannot Resist

much fun he could have with the rest of the crowd. After all, that's his field of research. Maybe he doesn't have the courage."

. . . .

Exchange students Andrés from Colombia and Kurt from Germany met at the University of the Balearic Islands on Majorca. During their nightly outings Kurt preferred to watch from the bar, beer in hand, while Andrés would take to the floor. Salsa, samba, mambo, or tango—nothing could stop the Colombian. I met the two of them in a dance hall on Palma's waterfront promenade. The way Kurt stood around looking cool was familiar: Germany and Denmark take top positions in the ranking of "men at the bar." I asked Andrés about his dance secret. "In Colombia, people think you're a complete idiot if you can't dance. No girl would ever look at you. And if you want to make friends and get invited to parties, you'd better learn to dance. Otherwise, you're on your own."

According to a survey in *Men's Health* magazine, 90 percent of women but only 10 percent of men in Northern European countries like to dance. You can observe this mismatch in many dance classes, where girls have long taken to dancing with girls because too few boys show up. As a result, not many boys learn to dance. And this, in turn, is the reason men don't dance: most claim they're bad at it. Going around in circles! Many men seem anxious about embarrassing themselves instead of appearing to be in control. Rather than move, they clutch their beer glasses and watch others have a good time. Many boys find dancing extremely uncool. The reasons seem to be complex. But it's a known fact, dear men, that women find men who dance very attractive. Taking dance lessons is worth it! Afraid of embarrassing yourself? Start with a beginners' class where *everyone* is a beginner. Your female partner will be grateful to have a male partner to dance with, and everyone else will be so caught up in worrying about their own feet that no one will watch you. Before you know it, you'll be able to dance.

When I met Kurt and Andrés again six months later at the farewell party to mark the end of their exchange year, Kurt had graduated with honors in swinging his hips. His Latin American buddy gave me a broad grin when he noticed my surprise: "I put the fear of God into him that he wouldn't find a girlfriend. In tribal cultures all over the world, a man first has to show off with a dance before a lady would do as much as glance at him. You girls only have yourselves to blame. You need to be more picky when you choose your partners. If you accept men who don't dance, you disrupt the laws of nature!" He winked at me. "If a man wants a woman, he first has to slay the beast of fear, including the fear of looking ridiculous. If a guy can't even face his fear of dancing, what else can't he do?"

Apparently, Colombians don't mince words, and if it's more embarrassing *not* to dance, people will dance. A Japanese saying has a similar message: "We are fools whether we dance or not, so we might as well do it."

• • • • • • • • • • •

If you ask me about the best present I have ever given anyone, I'll tell you, "I got my friend to dance." In fact, several of my friends, both male and female, have thanked me for passing on my passion for dance to them. It allowed them to find friends for life and introduced them to a lifelong hobby.

But not all of us have a friend who drags us onto the dance floor. And so we squat on our barstools, drink in hand, and marvel at the dancers. They are laughing, enjoying themselves, they move and flirt—while we sit at the bar and think, "I could never do that." However, very few people are actually incapable of learning to dance: only about 1.5 percent of the population, as it turns out. The probability that you are among them is truly miniscule. In the research literature, "Mattieu," as he is called, is known for belonging to this group. Although his ears are perfectly fine, he's incapable of recognizing any sequence of sounds or any rhythm, let alone repeating a melody he has heard. For the most part, he cannot even distinguish between two different melodies. "Most people can hear wrong notes. I can't," Mattieu admits. "To me, everything sounds alike." This congenital disorder is called "amusia." People suffering from amusia can neither classify rhythms nor match their body movements to a given rhythm. They can follow a metronome without problem, but not a rhythm inside a melody. Rather than being color-blind, they are music-blind—and hence dance-blind. People like Mattieu perceive any style of music as a kind of hammering sound or as a random, nerve-racking, and sometimes actually painful racket.

Mattieu, though, is the exception. The majority of people are not dance-blind. Rather, it is our insecurities and self-consciousness that keep so many of us from getting out on the floor. Self-consciousness is not an emotion we are born with. Last year, Laura, the daughter of one of my colleagues, still danced freely and unselfconsciously to the guitar in her summer camp. This summer, she stood stiff as a rod at the edge of the group; wild horses could not have made her join the others.

"But last year you really enjoyed dancing," her mother said, at a loss, as she tried to shoo her daughter from her side and onto the dance floor. What had happened?

Self-consciousness is an innate ability—sooner or later we all feel self-conscious—but it doesn't come into play until we are old enough to be able to see ourselves through the eyes of others. The moment children stop dancing spontaneously is the moment they begin to perceive themselves as social beings among other social beings. Between the ages of five and seven, they suddenly begin to care about how other people see them. That is when feelings of self-consciousness emerge, because we start to compare who we are with who we would like to be. And if those two images are not congruent, or if we at least believe them not to be congruent, we start to feel self-conscious. We reach the climax of that experience during puberty, when all eyes are on us. All eyes—really, every single pair of them—keep watching us, and only us: the bus driver, the neighbor, the friendly boy in tenth grade, and that girl who never talks to me.

Of course, that is not really the case, but during puberty different areas of our brain develop at different speeds, and this immature brain makes us feel weird. In these years we are one big construction site, both physically and psychologically, and hence very self-absorbed. The focus on ourselves makes us feel as if we are the center of the world, watched and judged by everyone else.

And that's not the end of it. During puberty we also believe that our thoughts, feelings, and fears are entirely obvious to others. We feel as if we live in a fishbowl. Depending on our personality, we may enjoy that feeling or wish we could hide all day long. And with this (mistaken) belief that we are the center of the world, we suddenly find ourselves on the dance floor. Children's birthday parties give way to teenagers' parties,

Solo Dance—Rhythm I Cannot Resist

where the more courageous among us start putting on music. Prom is approaching. While only yesterday we were toddlers with rabbit ears who danced cheerfully and without a care in the world, now we feel like gangly giraffes helplessly teetering on overly long legs at the edge of the dance floor. During puberty, when hormones have turned our face into a cratered landscape and even the eyes of our neighbor's dog make us feel self-conscious, just walking without stumbling can feel like a challenge. And in such a state we are suddenly supposed to float with grace or move with sexy self-confidence in front of the watchful eyes of others? That's bound to turn out badly!

The fact that moving in front of others becomes stressful is part of the completely natural process of growing up. Even in countries where dancing is part of everyday living, as in Andrés's homeland of Colombia, children start to feel self-conscious around the age of five. Initially, these children feel as unsure of themselves as their North American or European counterparts when they dance, but they accept the challenge because everyone else does. In Colombia, people dance everywhere and all the time: at company parties, birthdays, Christmas, and sometimes even just for the sake of dancing. In North America, though, it would be very strange if Grandma suddenly got up to do the twist in the middle of Thanksgiving dinner. At company parties we tend to stand around awkwardly and cling to our wine glass. Just imagine dancing salsa in the lab with your boss... No way! For most of us, dance is definitely not part of our daily lives.

 As a DJ, I have a few tricks up my sleeve to help people overcome their dancing inhibitions. For example, artificial fog helps to create a feeling of safety because people don't feel so watched. If I play songs for which there are currently popular moves (from "YMCA" to "Gangnam Style") and everyone repeats these moves, then most people will become less inhibited in their dancing.

From the point of view of developmental psychology, puberty may well be the worst time to start dancing; it can cause deep anxieties. Take our colleague Markus, who never wants to dance again. He is forty-five years old and has a family and a fantastic job. He loves to take his wife out for dinner, concerts, or to street parties. But when people start dancing, he flees. Just hearing a waltz makes him break into a sweat. His wife would love for them to take a dance class, because it's supposed to be so good for one's health. But she doesn't stand a chance. Markus won't go. Worse, he *really* can't. His legs turn to jelly at the mere thought. Markus is afraid—and his fear was caused by a traumatic experience at his prom. Proms in Germany are a bit different than in North America. Before the kids are set free to party and dance on their own, the parents are entertained with a performance by the graduating class. In Markus's case, that presentation included a waltz. It didn't go well. He tripped in his new shoes, which were too big and too slippery. Suddenly, he had to take a big sidestep to keep his balance, and he almost pushed his partner off the stage set up in the school's assembly hall. Red as a beet, drenched in sweat, and with a racing heart, he finished the dance with his partner and then fled from the stage. Markus had embarrassed himself in front of his friends, all the parents, and worst of all, his dance partner. Today, none of his classmates or family members have any recollection of that moment, and Markus himself doesn't even remember the girl's first name. Nevertheless, the feeling of having made such a fool of himself is deeply etched into his memory.

Unfortunately, unpleasant experiences we suffer during puberty are stored in our brains as memories and can take our breath away even when we have a fully matured adult brain at our disposal. They can let us, like Markus, cling to the rock-solid belief "I can't dance." The fear of being stared at and judged when we are on the dance floor accompanies us into adulthood.

Often, we believe that the outside world is somehow aware of the insecurities we feel. In fact, experiments have shown that research participants are convinced others can see their fears and insecurities much more clearly than is actually the case. When researchers ask observers what they see, and compare that to what the participants *think* is seen, they find a big discrepancy. We believe others assess and judge us more harshly than they do, and we may even believe that our desires and preferences are completely obvious to the outside world. However, if we were able to see the people around us more objectively, we would find that most don't even look at us as they go about their own business (grappling with their own fears and insecurities).

Maybe now you can see how dancing can actually make us more confident. People who dance have faced up to the risk of embarrassing themselves, have listened to their own heart, don't give two figs about other people's thoughts, and instead surrender to the pleasure of dance. People who dance are strong. As the American dancer Agnes de Mille put it, "To dance is to be out of yourself. Larger, more beautiful, more powerful. This is power, it is glory on earth, and it is yours for the taking."

"But I don't know the steps! How's dancing going to look when it's not done correctly?" We often hear that question when we share our enthusiasm about our hobby. But dancing is not about doing things right or wrong; it's about doing what *feels* right. Of course, professional dancers must move according to their choreographer's plan, but the most important thing for any dancer is the expression, the passion. Unless you perform on a stage, it's irrelevant whether anyone watches; what *is* relevant is what you are experiencing in the process.

> A friend of mine met her future husband in a salsa class. They fell head over heels in love during their first dance. He soon proposed, and she accepted. When they planned their wedding celebration, they decided to perform a dance for their guests; after all, dancing was how they'd met. They practiced hard; they wanted their performance to be the highlight of the evening. But the whole thing turned into a complete fiasco. Their first attempt ended when the bride tripped on the carpet. In their second attempt, the bride crashed into the DJ's table because the bridegroom failed to catch her elegantly after a turn. And in their last attempt they were both so nervous that they fell out of step and danced more against than with the music. What happened? In class they'd always been such a harmonious pair. But there they had danced for their own pleasure, had focused on themselves, and had therefore been able to move in perfect coordination. There, it was all about the dance and not about the recognition they might get from others. It is only when the world around you vanishes and no longer matters that you are truly dancing.

We are no longer adolescents; we should know we're not the center of the world. Not even close. If you scan the dance floor, you will notice that your fellow dancers don't likely care one bit about how you move. And all those people standing on the sidelines staring at the dance floor are probably there for one reason only: because they're convinced they can't dance. Or they don't have the courage to dance. You are already well ahead of them. You have the courage, you're having fun, and you're in a good mood.

THE PONY FARM DANCE

By now most of our colleagues have made their way onto the dance floor. It's interesting to watch if and how the way people move when they dance reveals something about who they are during the day. There's the cool guy with minimal movements, hands in his pockets; the quiet one with closed eyes and a blissed-out smile who has left all earthly concerns behind. The macho dude with his need for space; the sunny woman whose arms and legs swing freely; and the wild one who loves to let her hair fly. When we dance, we express ourselves and show our emotions. Without saying a single word.

Our bodies communicate information all the time. Consider just a single arm movement, for example. Bend your arm at a ninety-degree angle from your elbow, palm facing upward. Interpreted one way, this gesture suggests that you are about to receive something very special and hold it in your palm. However, you can inject a bit of rage into this same position by thrusting your open palm upward in an accusatory gesture, or you can raise your arm with its empty palm to indicate a lack of energy. A body is never mute. Depending on how we stand or walk, whether we cross our arms or allow them to hang at our sides, or whether we stand, legs apart, or menacingly stride up to someone—our body is the direct mouthpiece of our emotions. This nonverbal body language typically communicates below the level of our awareness and therefore can reveal more about us than we intend to express with words. It gives others an honest and authentic insight into our emotions.

Emotions are our reactions to events. Researchers have classified human emotions into six categories: happiness, sadness, anger, fear, disgust, and surprise. Through these emotions we react to our social environment and appraise for others as well as for ourselves what is happening at any given moment. We perceive through our senses what is happening around us, and when the nerve impulses reach our brain, we translate them into an interpretation of the events. Emotions are the result of our body's reaction—the firing of nerve cells and the stimulation of hormones and other chemical messengers. Different emotions lead to different physical responses and hence feel different in our body. There is no such thing as an "anger hormone" or "happiness hormone"; instead, the coexistence and interconnectedness of our emotions works like an audio mixer console. Much depends on the combination and context. Anger releases testosterone, but so does happiness. Fear, surprise, even intense happiness release adrenaline. And if we are newly in love, serotonin, noradrenaline, and dopamine make us see the world through rose-colored glasses.

But it is not necessary to receive a stimulus from the outside world in order for an emotion to be triggered; our own thoughts can do the trick. You probably know what it feels like to daydream and suddenly have your heart race or your hands sweat, as if what you merely imagined had actually happened. Imagined emotions have the same biological effect on our brain as lived emotions. They activate our biological mixer console and blend hormones and neurotransmitters. This is actually an amazing thing, since it gives us a great deal of control when it comes to manipulating our emotional inner world. If we think of beautiful things or relive positive experiences in our mind, our mixer will create the perfect "sound" for happiness. We can even reinforce the emotion with our body language by "trying on" the emotion. A well-known trick to overcome a bad mood is to smile. The mimicking of happiness sends a "good mood" signal to our brain, which in turn releases the respective hormones—and bingo, our actual mood has improved.

Hardly any other sport is as suitable as dance for the "trying on" of emotions. Choreographer and dancer Ted Shawn, one of America's modern-dance pioneers, was convinced that we can express our emotions much better through dance than through language.

In 2015, neuroscientist and dancer Asaf Bachrach from the University of Paris was able to demonstrate that audiences of particular dance performances physically share the emotions of the dancers. The viewers of a modern-dance performance had their breath measured, and it was revealed that they were breathing in time with the dancers. Even more surprising, the researchers discovered that the more the breathing of the viewers and dancers was synchronized, the more the viewers enjoyed the performance. Modern dance, with its greater freedom of emotional expression, seemed to invite viewers perhaps more than other dance forms to experience for themselves the emotions presented by the dance.

Professional dancers use their art as language, and they master the secrets of nonverbal communication. They are especially good at expressing emotions through movement. The expression we observe in a dancer is often the result of that dancer's authentic inner feelings, even if they are dancing steps prescribed by choreography. One can dance the same step sequence happily, sadly, or angrily. World-famous choreographer Martha Graham used to emphasize that outstanding dancers are not the result of their technique or perfectionism but of their passion.

Professional dancers draw on a variety of tricks to instil emotion in their dance. They work on remembering certain emotionally charged moments in their life, and then slip those emotions on like a piece of clothing and express them in their dance. It's something anyone can practice. Give it a try.

Sit in a comfortable seat and imagine opening a letter you just picked up from your mailbox. The letter is from a dear friend you haven't seen for a very long time. Hearing from him makes you incredibly happy, and your attention is totally focused as you begin to read. Your friend tells you of a recent and incredible stroke of good luck. You are surprised and happy for him, because this good fortune led him to a fantastic new job. Your friend shares a few anecdotes from work, and also refers to experiences you shared many years ago. They make you giggle, just as they did then. But then you get a shock: without any warning, he gives you extremely sad news about some people you both know. At some point, your sadness turns to anger. The details your friend includes show that the disaster only happened because somebody deliberately did something wrong. You are furious at the injustice. You're also disgusted! Your friend continues by describing how he can imagine that you are now beside yourself with anger, and he asks that you do what he himself did: shrug it off and move on. He then gives the whole story a bit of a comical twist, which makes you smile again. You remember that he always managed to do that, regardless of what was going on in your lives. He closes his letter by expressing his hope to see you again soon. You take a deep breath. You hope so, too!

In the drawings above, you see a person whose facial expressions show the different emotions involved in the exercise. You can imitate those emotional expressions, but the results would be "empty" and faked, not authentic. If, however, you walk yourself through the letter exercise and, in the process, generate genuine feelings in yourself, your expressions will in turn become genuine; only then will muscles we cannot consciously train become involved.

In dance, emotions expressed by movement are very closely connected to music. In 2014, I worked on an interesting experiment. One day, as I stood with a colleague on the fourth floor of our university building, I had an idea. It was during a break, and the noise of hundreds of students in the seminar rooms had been deafening. I was listening to classical music on my headphones to block out the noise and was watching the movements of the crowds of students on the ground floor. As the music was playing in my ears, the way the students flowed in lines and small groups from one side of the lobby to the other looked like a ballet. A bit like *Swan Lake,* I thought, and actually quite beautiful. I shared my impression with my colleague, and he, too, watched the scene for a moment. Because the music had been a random choice and had absolutely nothing to do with those swarms of students, we wondered whether it was perhaps completely irrelevant what kind of music is played when we see movements. Perhaps we simply find any combination of music and movement beautiful? If you can, open side-by-side windows on your internet browser. In one, find the YouTube video with Herbert Grönemeyer's "Letzter Tag," in which world-famous ballerina Polina Semionova dances solo on a stage. In the second window, open any music video you like: Metallica, Eminem, or whatever comes to mind. Mute the sound of the Grönemeyer video so you can watch Semionova dance while listening to the other video's music. See? Astounding, isn't it? Semionova's movements fit the music.

Our brain's neural mechanisms allow us to create a connection where none exists. If a movement occurs at the same time as a melody, our brain creates the illusion that they are synchronous and match each other. It's a clever party trick, for certain, but does our brain really see these unconnected things in the same way it does music and movements that have been specifically designed for each other? Does any mix of dance and music spark joy? If so, why do we need artists to create dance choreographies?

In the experiment we designed based on these questions, the participants were shown a series of short video clips with ballet sequences. They were asked to evaluate the clips and to determine the emotion expressed in each. While they were watching the dancers, the participants also listened on headphones to different pieces of music we played for them. We asked them to ignore the music they heard and to concentrate solely on the dance movements they saw. Those dance clips showed cheerful as well as melancholy sequences: sad ones from *Swan Lake* and high-spirited ones from *The Nutcracker*. The music on the participants' headphones was at times cheerful and at other times sad, and sometimes we played none at all. Meanwhile, we were measuring their skin-conductance levels, a method used to check whether arousal makes their hands sweat. Not every emotion causes our hands to sweat to the point where we can feel it, but finger electrodes can pick up even minimal changes. Skin-conductance measurements are hence an indirect means of measuring what the brain is doing at any given moment and whether we are emotionally aroused.

The results were astonishing. The participants' brains couldn't help but register whether the dance movements and music agreed in their emotional expression. Without consciously paying attention to the music, the participants got emotionally aroused only when music and dance expressed the same emotion; that is, when a joyous dance was paired with joyous music or a sad dance with sad music.

It's a fact of nature that we can express our emotions through movement. What do people do in their happiest moments? And how do we know we are happy? We have a radiant smile, ringing laughter, or eyes shining with happiness. But there's something else we do unthinkingly: we move. When children are happy, they bounce up and down; when we celebrate a success, we throw our arms up into the air; and when we are in a gloomy mood, we let our shoulders and head droop. When we meet friends we haven't seen for a long time and have missed, we hug them, and we sway together as if to some inaudible music. Dance is part of any joyous celebration: in almost every culture people dance at weddings. Germans, Italians, Greeks, Russians, South Asians, Arabs, and even people from more reserved cultures dance at family celebrations. Our movements—especially dance—let those around us learn much about how we are feeling in that moment. And we can use dancing to deliberately express how we feel at a particular moment.

DANCING IS THE BEST MEDICINE

When I was about five years old, Simon & Garfunkel's "Bridge Over Troubled Water" was my absolute favorite. I didn't understand all of the English lyrics, but I grasped the feeling in the music: it was about longing. Whenever I heard the song, I'd feel the same way I felt when I was missing Sofus, my favorite pony at the farm. And that longing was a feeling I could dance. I would raise my arms, sometimes high over my head, sometimes sideways, as if I were trying to grab something that was unfortunately beyond my reach. The song contains long, flowing sounds—today I recognize them as notes played on the violin—that one could gracefully glide on, especially in a circle with both arms stretched out like an eagle flying across the water to the pony farm (we had to drive over a long bridge across a canal to get there). "What's Julia doing?" my grandma asked one day. "It's the pony farm dance," my mother explained. "When Julia misses Sofus too badly, she puts on the Simon & Garfunkel record and dances this dance. She feels it shortens the wait before she gets to go there again."

* * *

Night has fallen in Greece. We've talked for hours and have discussed many aspects of dancing. Isn't it amazing that babies and toddlers already recognize rhythm? And that every dance can reveal something about our emotional life to those who are watching? The topic is fascinating, and a single evening is not nearly long enough for us to exchange all our thoughts and ideas. But we've said enough for one night—it's time to dance!

2

PARTNER DANCE– DO YOU SPEAK DANCE?

....
Dance is Esperanto with your whole body.
FRED ASTAIRE
....

T he second day of the conference is packed with interesting topics: Why do we experience empathy? What connects us to other people? To our partner? We examine all of these questions from a scientific perspective: What is happening in our brains when we feel empathy or connect with others? A great deal of new information is shared, and many topics are rigorously debated. The evening is thus a welcome change, a chance to hang out at the bar and enjoy the setting sun. The DJ who'd kept us moving into the wee hours the night before *is not at his station. Instead, a band plays Latin American dance music, and a few couples are dancing the rumba. One older couple attracts everyone's attention. They move in great harmony and smile at each other often. Sometimes their movements are almost synchronous; at other times they appear to be having a conversation, where one movement responds to the other.*

SOMETHING IN THE WAY SHE MOVES

In the 1970s, British director Peter Brook took a multicultural troupe of actors—including a very young Helen Mirren—to the Sahara Desert to conduct a daring experiment. Brook wanted to find out to what degree a stage play could be communicated nonverbally, using only the body. The troupe performed in villages and towns and often in front of audiences that had never seen a stage play performed this way. Brook was deeply convinced that we can express our commonalities through facial expressions, gestures, and other movements, regardless of the language we speak or the culture from which we come. His core idea was that movement is the foundation of all communication.

Humans have the most complex brain and the most differentiated repertoire of movements of any species. We employ movement in all areas of life and ultimately communicate through movement. We have a highly complex vocal apparatus, elaborate facial expressions created by the movements of countless muscles, and are capable of showing fellow human beings our feelings and intentions through gestures. Movements allow us to communicate and understand each other. Some researchers even believe that we have brains for the sole purpose of controlling movement, and that in turn our brains have evolved so much only because we move. Neuroscientist

Daniel Wolpert of Cambridge University illustrates this theory vividly in a TED Talk that uses the example of the sea squirt. As a larva, the sea squirt is free to float in the ocean. It has a nervous system and a primitive brain. Once the sea squirt is mature, however, it looks for something to attach itself to for the rest of its life. Over time, it will then digest its brain, because a brain is apparently no longer necessary now that it has "settled down." Wolpert likes to joke that sea squirts are like professors: once tenured and no longer forced to move, they devour their brains and become mindless.

Being able to move, to interact with our environment, is an important skill. In its first years, a baby doesn't have a verbal language but learns through movement. Initially its movements look entirely random, but they soon become more directed. The baby learns to lift its head, to grasp for things, to roll over—and through these movements its brain discovers the world. With time, the baby will connect specific movements with visual and auditory stimuli. For example, the baby may reach for a rattle. "Would you like that?" the mother may ask with a smile as she places the rattle into the baby's hand. In the child's brain, the nerve cells responsible for visual, auditory, and movement signals will fire together, in the same rhythm. Language builds on action. Movement leads a child to watch the mother's response and increasingly learn to understand the concept of "have" until it can apply it him- or herself—at which point the child's demanding "Have it!" will ring out ever more often. The interaction of moving, seeing, and hearing lends meaning to words and hence to language. Slow and fast, near and far, but also round and square, soft and hard—ultimately, a toddler discovers the meaning of all these words through its movements.

As a result, we observe in children a relationship between fine motor skills and linguistic facility. But that's true not only for children. Through fascinating experiments, neuroscientists Friedemann Pulvermüller and Arthur Glenberg have demonstrated that preventing people from making certain movements slows down their word recognition. In one study, participants with their hands tied behind their backs had a harder time recognizing words whose meaning involves the use of one's hands. We often use gestures when we speak. We indicate whether something is large or small, emphasize helplessness by hunching our shoulders, or reluctance by crossing our arms. That the gestures that accompany our words are important can be seen particularly well in situations where we cannot be sure our counterpart can really understand what we are saying. Whether we whisper during a show in the theater, try to tell someone something in a crowded bar, or attempt to communicate with someone who isn't fluent in our language—in any such situation we will intuitively use more gestures. We are also very skilled in interpreting other people's gestures. Research has shown that this interpretation activates the same brain systems as the interpretation of language. In fact, everything can be told with movements alone—for instance, in sign languages that are used in communities of the deaf—without the need for any verbal language at all. Everyday conversations, heated debates of quantum theory, as well as beautiful poems work as well in sign language as in verbal language.

Scientists at the Max Planck Institute for Human Cognition and Brain Sciences in Leipzig discovered that vocabulary is easier to learn when we connect new words to movements during the learning process. The effect of this strategy could even be observed/detected in the research participants' brain activity, as Katja Mayer and her colleagues were able to demonstrate in 2015 by using fMRI (functional magnetic resonance imagery): if, after learning new words, a study participant translated a word together with a gesture, the areas of their brain responsible for movement were activated. Should we perhaps dance our vocabulary?

The University of Sheffield's Steven Brown and his colleagues at several other institutions had participants in an experiment practice Argentine tango regularly over an extended period of time. After a few weeks, the researchers showed the participants dance videos while measuring their brain activity during a CT scan. They found that the neural networks typically responsible for processing language were now also active while the participants watched the dance moves they had learned—as if the moves they were watching were speaking to them! As far as the brain is concerned, dance appears to be a language.

Science seems to confirm what tap dancer and actor Fred Astaire instinctively knew: "Dance is Esperanto with your whole body." And once we know the "vocabulary," or the basic steps, of a particular style of dancing, we can form sentences and tell whole stories with our movements without having to think about how these movements are put together. We can "speak" that dance language without having to repeat the same step sequence, just as we combine words in a language in new combinations to express meaning and to form sentences we have never uttered before. Similarly, a dance comes alive when the dancer no longer just does the steps but *dances*.

MIRROR DANCING

Earlier today we listened to a lecture about social mimicry. That's scientist-speak for the way we unconsciously adjust to the movements of the person opposite us—but only if we like them! Experiments have shown that social mimicry only

happens between individuals who like each other—and that individuals who don't know each other may start liking people who "social mimic" them. Examples of social mimicry can easily be found on the hotel's dance floor. But not all dancers move as harmoniously as the older couple we were watching earlier in the evening. It's interesting how much we can learn about people we've never seen before simply by watching them. With one couple, for example, there is zero mimicry at play. They don't seem to like each other one bit; in fact, their dance looks more like a battle.

"As if they've been forced to dance with each other! They look like competitors, or even enemies."

"Oh well... maybe they've been married to each other for twenty years?!"

Next time you're out on a busy street, take a good look at your fellow human beings. You can get a pretty good idea just from someone's gait whether they're on top of the world or miserable. Do they walk with drooping shoulders and steps as heavy as if they weighed a ton? Or do they move down the sidewalk with a spring in their step? Our mood—be it sad or happy—will be reflected in our movements. The idea that we can understand the emotions of others by embodying their movements is not new. As early as 1759, British philosopher Adam Smith spoke of "reflexive imitation." Charles Darwin later called it "motor sympathy," and in the 1960s social psychologist Gordon Allport wrote about "objective motor mimicry."

The pony farm dance allowed young Julia to express just how much she missed her little white pony, even when she lacked the words. Her dance made visible what was otherwise invisible: her feelings. Julia's mother understood her daughter's wistful movements and was able to "translate" them for Julia's grandmother. In fact, studies show four-year-olds expressing their fear, anger, sadness, or happiness through dance without even being aware of their actions. The children simply feel and dance their emotions. In addition, children who merely watch others dance also understand their emotions. Children as young as five can name the emotions expressed in a dance.

Dance training helps to further develop this skill. The more a person learns to translate their own emotions into movement, the more capable that person becomes of appreciating the emotions in other people's movements. In a study conducted at City, University of London, we showed short dance videos to professional ballet dancers and to a control group of people with no dance experience. The video clips were muted, black and white, only a few seconds long, and had the dancers' faces blurred so that no facial expressions could be discerned. All participants were asked to appraise the emotions that the dancers were expressing in the videos merely by studying their body movements. Were they happy or sad? Again, we measured skin-conductance levels to gauge the physiological reactions of our participants while watching the videos. Both the dancers and the non-dancers correctly classified the movements as "happy" or "sad" when the movements were indeed taken from happy or sad moments in the choreography. However, the participants who were dancers were much more sensitive to the emotional content than the control group of non-dancers. Are dancers not only experts in dance but also in emotions? Our study suggests that we can improve our conscious perception of other people's body language through dance training.

Dance communicates many different messages, and we can have real conversations solely by using our body language even if a dance stems from a different culture. In 2000, psychologist

Ahalya Hejmadi of the University of Maryland and her colleagues studied whether our understanding of the emotional expressiveness of dance is universal. To do so, they drew on Indian dance because its ways of expressing emotions through movements are precisely prescribed by a two-thousand-year-old treatise on Indian dance, the *Natya Shastra*. Each movement follows a strict protocol. The emotional expression involves the whole body, including facial expressions and hand gestures. The researchers filmed a dancer who expressed ten classic Indian dance emotions and showed the videos both to Americans and Indians. The participants were asked to identify the emotions expressed by the dancer. Surprisingly, both the Indians and the Americans were able to perfectly identify the emotions expressed in the dances.

But how does that work? How is it that we understand the body language of our fellow human beings even if we belong to different cultures? And how are we able to do so already in childhood, without ever having been taught? What happens in our brain that makes us associate emotions so precisely with certain movements and lets us read the emotional quality of those movements?

As part of the research for my dissertation, I explored whether our first impression of a person is shaped by that person's gestures and movements. I wanted to find out whether someone's "personality" permeates their movements and whether others can recognize it. To answer those questions, I asked participants to perform a series of movements: walking, running, and jumping, as well as more complex ones such as playing table tennis or dancing. For dancing I used two subcategories: freestyle dancing or dancing to a set choreography. For the latter, I picked the "Macarena" because everyone knows the song and how to dance to it. I filmed all the movements and then projected them onto neutral avatars so extraneous details such as individual looks wouldn't interfere with the participants' appraisal of the movements. Then I asked the participants—who could now be seen as avatars in those short clips—to assess personality traits, such as how cooperative and trustworthy the avatars appeared to be. A separate group of participants was similarly asked to assess the avatars. The results allowed me to compare the (self) assessments of the first group with the assessments of the second group. I was convinced I would see the strongest agreement with regard to the freestyle dance movements; after all, in freestyle everyone finds their own style, making it more likely that personality traits would be revealed. Or so I thought. The results came as a surprise. The participants agreed far more in their personality assessments for the "Macarena" than for the freestyle dancing. The set choreography appeared to allow the participants in the second group to imagine doing the movements themselves—to essentially dance in the dancers' shoes. Because they had at some point in their lives made those very movements themselves, they were better qualified to assess them. In a way, they were looking at a mirror.

We owe the scientific explanation for this phenomenon to a chance discovery in 1992 that turned the world of neuroscience upside down. Professors Rizzolatti, Gallese, Fogassi, and di Pellegrino of the University of Parma were working on identifying the nerve cells that are activated when a macaque reaches for an object. They had trained the monkey to pick up different objects, move them a short distance across a table, and put them down again. The monkey was connected to a measuring device that was specifically designed to pick up activity in certain nerve cells in the brain. The experiment began with the monkey being offered a tray with various objects. The animal took the objects and moved them one after the other across the table. Meanwhile, the researchers were able to measure distinct activity in those areas of the brain that control movement. Once the macaque had moved all the objects, the research assistant prepared the next tray, which was to include three additional objects from a cupboard. As the assistant was reaching for these objects, nerve cells in the monkey's brain suddenly started to fire in the same way as they'd done while the monkey itself had reached for the objects. Perplexed, the assistant stared at the measuring device, and then looked at the monkey, who was watching him attentively. The monkey was sitting quietly, and the measuring device seemed perfectly okay. Had the animal moved? The assistant checked the monkey's paw. Had the little guy swiped something? It hadn't. Without taking his eyes off the monkey, the assistant once more moved his own hand slowly toward the object on the tray between him and the monkey. And again: the nerve cells in the monkey's brain fired when the assistant's hand closed around a nut. They reacted even though the monkey had done nothing but watch the assistant's movement. The monkey's brain was, in a way, mirroring the assistant's movement. You can imagine the excitement this discovery caused among the researchers!

    The researchers called the nerve cells that fired when the monkey observed a movement "mirror neurons." Since then, these mirror neurons have been the subject of numerous studies. Yet we are still far from being able to determine conclusively what these neurons are capable of and what they are good for. Researchers assume mirror neurons are involved in our understanding of other people's body language and emotions—and maybe in empathy: your body language shows me how you feel, and because my brain mirrors your emotional state in my body, I understand you. We've all experienced this process at some time or another. Give it a try: Put a slice of lemon in your mouth and watch the person opposite you. They can't help but make a face, as if they had just bitten off a piece of lemon themselves. In a way, we absorb other people's feelings. And this happens from the moment we are born because mirror neurons arrive as part of the basics of our brain and are fully developed by the end of our fourth year. Mirror neurons activate the same areas of our brain that would be activated if we made those movements ourselves. We understand the body language of other people because our brain knows automatically what it feels like when we make the same movements.

English neuroscientist Patrick Haggard specializes in how the brain controls complex movements. A lover of ballet, Haggard wondered what effect an art of movement as sophisticated as ballet has on the areas of our brain that control movement. He would often reach out to ballet dancers to talk, and during these conversations he noticed that they appeared to see much more than he did when watching other dancers. He and his then PhD student Beatriz Calvo-Merino filmed both ballet dancers and capoeira dancers doing their respective typical dance movements. In a second step, the researchers then showed the video clips to groups of ballet and capoeira dancers, as well as to people with no dance experience (non-dancers), while measuring their brain activity with fMRI. They found that the ballet and capoeira dancers showed distinctly more activity in important mirror neuron systems of their brain while watching their own style. No such difference in brain activity was found when the people without dance training watched the videos of the two different dance styles.

This study showed that dance training allows dancers to perceive movements differently than those without dance expertise. In fact, the evidence suggested that professional dancers seem to "dance along" in their mind, whether they are dancing themselves or merely watching others dance. Following in the footsteps of this seminal study, numerous others have shown the enticing effect of dance expertise on the brain.

The exciting thing about mirroring is this: Thanks to our brain's ability to mirror the activities of the person opposite us, it can anticipate what the other person will do next. When we see someone "give" something, the "take" action is, for our brain, already encoded. We understand others by simulating their actions in our own brains. And that enables us not only to understand each other but also to work together, to cooperate.

In our social interactions, we go even further. If two people get along well, their facial muscles and hence the movements in their faces align with each other's without either person being aware of it: if one of them looks surprised, the other one does too; if one smiles, the other will as well. If facial muscles have been paralyzed, say with Botox, and this alignment is prevented, we have a harder time understanding the other person's feelings. Studies such as the one done in 2011 by David Neal and Tanya Chartrand at the University of Southern California have shown that Botox injections even flatten the emotional life of the recipient.

Social mimicry is, by the way, a very effective flirting strategy if used deliberately. If we imitate the body language of another person, they will unconsciously sense a connection, which in turn will make them like us more. If we watch video recordings of such encounters with the sound muted, the pair's movements tend to resemble a perfectly choreographed dance.

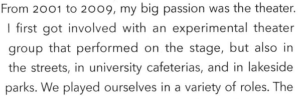

From 2001 to 2009, my big passion was the theater. I first got involved with an experimental theater group that performed on the stage, but also in the streets, in university cafeterias, and in lakeside parks. We played ourselves in a variety of roles. The American poet Walt Whitman once famously wrote, "I am large. I contain multitudes." We came to know ourselves through theater.

I met many dancers in that group. Gerhard, our director at the time, experimented a lot with dance. One of our exercises was the "mirror game." We would gaze at a partner and try to imitate their movements in a slow, steady pattern to achieve synchrony—like watching ourselves in the mirror! Bernhard, a professional dancer, guided us in these exercises, and after a while we did indeed manage to synchronize our movements. The most striking aspect of the exercise was that we learned to imitate each other's movements without first determining who would lead and who would follow. Afterward, we would feel very close to each other, as if this dance had allowed us to connect through both body and soul.

Later, as a neuroscientist, I realized that we had truly synchronized our brain waves in those moments. Our brains did the work of simulation for both ourselves and the other person at the same time, as if we were one and the same. Our brain performs this co-activation "magic" not only when performing a "mirror" dance but also—and rather frequently—in our day-to-day lives. This automatic mimicry process allows us to "read" the movements of others so we can empathize with them and ultimately understand their emotions and intentions.

This "mirror game" is one of the best-known acting exercises for improvisation. Movements are perfectly synchronized thanks to our brain's combination of mirroring and anticipating movements. It is this very mechanism that makes partner dance possible. One of the dancers leads, the other follows, and eventually it all becomes one unified, fluid movement—thanks to our brain's mirroring capacity.

This effect is not only at play when we dance but also when we make music together or listen to each other. Uri Hasson of Stanford University researched the brain activity of people who read stories to each other. His experiments showed that the listener's brain activity mirrored that of the reader.

Swing dancers have a special kind of dance competition, the so-called Jack & Jill. Leaders and followers are drawn by lots, ensuring that dancers who don't know each other are paired up. The first steps with an unfamiliar partner tend to be somewhat awkward and hesitant, but soon enough they "click," and they almost automatically synchronize their moves. It's great fun, and there's always much laughter. After a dance, a (verbal) conversation flows much more easily, because the partners' brains have already harmonized on the dance floor—dancing together creates a bond.

MAY I HAVE THIS DANCE?

While we watch the dancers, we play a guessing game. Who has known whom for a while, and who has just met? If we look carefully, it's not hard to figure it out. And we can even catch a glimpse of the future.

"I bet those two will have hooked up by the end of the conference!" We are following a pair of colleagues who are dancing salsa. She gave an interesting paper this morning on mirror neurons, and he kept the discussion that followed lively by asking lots of questions. What had appeared to be professional interest in the context of scientific debate looks quite different on the dance floor. You can positively sense the air vibrating between them. More than just mirror neurons seems active here!

During summers in Berlin, dance parties regularly take place in a tent in the middle of a forest. Projectors cast images of wildly dancing people from the legendary 1970s TV series *Soul Train* onto the sides of the tent. The images are accompanied by music from the same era. It is a late summer night, just after midnight. The air is warm and humid, and the faces are glowing. It is nearly impossible to resist the pull of the music, to not get caught up in the energy of the dancing crowd. Surrounded by moving bodies, two dancers exchange a glance. Very slowly, they approach each other—step by step, beat by beat, floating on the melody of the music until they are together on the floor. They don't know each other's names, but they dance. Their eyes lock, and for this moment, time stands still. Two dance as one, in perfect synchrony. Sometimes he leans over her; at other times, she leans over him. Their dance comes in waves. He makes a move; she receives and mirrors it. She smiles; he smiles back. They dance for hours without exchanging a word. Later, outside the tent, drained but happy, the time has finally come to speak. Talking seems strangely unnecessary, though. They both feel as if they've known each other for a long time. No words are needed.

This is the story of how one of our friends met his partner. The evening happened a long time ago, and the two were a couple for years afterward.

We all remember couples from the great romantic movies: whether Scarlett and Rhett in *Gone With the Wind*, Tony and Maria in *West Side Story*, or Rose and Jack in *Titanic*—they all experience a crucial moment of coming together while they dance, and suddenly see their partner with new eyes.

What makes partner dancing so fascinating is the need for the dancers to synchronize their movements without using words—whether it's freestyle dancing or classical partner dance. On the dance floor, exciting things are at play in our body, mind, and brain. Dancing with a partner is like having a good conversation.

Partner Dance—Do You Speak Dance?

Our prehistoric ancestors may also have danced in couples, but the typical European partner dances are much more recent. During the Renaissance, in the fourteenth and fifteenth centuries, the European aristocracy developed specific social dances at the courts. Men and women would form lines facing each other, with a gap in between, and would exchange places and partners according to strict step sequences. You've probably watched costume dramas where men in powdered wigs line up and pass along ladies in fancy dresses with barely enough time in between steps to exchange a sentence. It was only during the Baroque era that real partner dancing began. It is said that King Louis XIV danced the first-ever minuet with his mistress in 1653. From then on, men and women danced together as couples. Yet those dances were still very different from the partner dances we know today. It was considered improper, for example, to touch the female dancer. If you absolutely had to, then only the hand please, for decency's sake! For a whole century, the minuet remained the fashionable dance *par excellence* and would soon be danced by the middle classes too. Simpler step sequences began to take hold, and normal folks weren't as prim about touching a waist or an arm. Nevertheless, when the waltz emerged on the eve of the French Revolution, it was an enormous scandal. This new dance allowed for physical contact, and dancers whirled so fast that a woman's ankles could be exposed! Not until the Congress of Vienna in 1814-15 did the waltz become truly acceptable in polite society, which explains why to this day it's called the "Viennese waltz." Ever since, partner dancing with physical contact and set step sequences has been the dominant dance style.

In many places, the Viennese waltz is still the traditional opening dance at a wedding reception. Because many couples don't know how to dance it, dance schools now offer crash courses for couples about to get married, which would have been unimaginable in the past. Everyone, including those who could barely afford to, used to take dance lessons—if for no other reason than that dance studios were the only place, in the era of single-sex schools, where young people of the same age and both sexes could be together in the same room.

With great enthusiasm my grandfather used to tell me how, as a young man, he would lug three milk churns around on Saturday mornings, delivering milk from the farm in order to earn his admission ticket for that evening's dance lesson. After the delivery, he would take a bath, polish his shoes, pull his Sunday best from the closet a day early (against his mother's protest), and walk the two miles to the dance school. It was the highlight of his week! That dance school still exists. For decades it has been owned by the same family and is an institution in its community. Generations have taken to its dance floor.

With the rise of gender equality in the 1960s, a result of the sexual revolution, partner dancing went out of fashion. People danced freestyle, mostly on their own. Partner dancing was considered old-fashioned and fuddy-duddy, as were the etiquette and protocol practiced at many dance schools. Smart studios began to offer courses that reflected new trends such as jazz dance and, later, break dance and hip-hop. But in the

1980s, movies such as *Dirty Dancing* rekindled the desire for partner dancing, and since then partner dancing has regained importance.

Partner dancing is an essential part of many social occasions. People waltz at weddings and groove to the blues at bars. Even teenagers attempt partner dancing at high-school dances and proms. In Europe, carnival season—a riotous celebration that takes place each year in the months before Lent—features countless dance parties, parades, and balls. The most famous of the balls, and always the season's highlight, is the Vienna Opera Ball, which takes place in the Vienna State Opera. Everybody who is anybody wants to be seen at the Opera Ball—and some seven thousand people (including staff) attend. Tickets for prestigious box seats start at 20,500 euros, or roughly US$22,000. The price of a small car for an evening of dancing! Nowhere else can people admire more beautiful ball gowns, so glossy magazines always cover the annual event.

Television, too, has contributed to the renewed popularity of classic partner dancing. Since its North American debut in 2005, *Dancing With the Stars* has been a perennial hit. In 2019, the finale of its twenty-eighth season was watched by 7.7 million people. The series' concept is simple: professional dancers pair with celebrities to compete in ballroom and Latin dancing. The television audience decides who will be eliminated and who may go on. The program's success comes from its mixture of grace and awkwardness, the pleasure of its music, its glamor, its specially designed costumes, and to some degree also from the *schadenfreude* viewers enjoy when dancers are eliminated. But most importantly, the show offers clear evidence that everyone—from politicians to "it girls" to sports stars—can learn to dance.

THE SEARCH FOR A PARTNER

While we're still discussing social mimicry and mirror neurons, things are heating up on the dance floor. Our two salsa-dancing colleagues are now glued together and swaying to their own rhythm, even though the band is playing a catchy samba. They don't seem to notice. His hands are on her hips, while she has closed her eyes and wrapped her arms around his neck. They move together, almost in synchrony, and although their movements are not quite in time with the music, they are a picture of harmony.

"Isn't it wonderful to get to know one another while dancing?"

While writing this book, we spoke to many people about dancing. It was fascinating that regardless of whom we were talking to—dance teachers, neuroscience professors, or seniors' home residents—most people knew at least one couple who had met on the dance floor. Dancing is indeed a wonderful way for singles to meet, and for couples to rekindle their romance. There are even couples therapists who use dance as an integral part of their therapeutic approach.

Perhaps you can recall a recent dance experience with your partner that was in no way "romantic" or harmonious: maybe you stepped on your partner's toes while he hissed a stressed "This way!" into your ear and yanked you around like a clumsy oaf. And partner dancing is supposed to be good for one's relationship? We can assure you: it really is. But the magic word is *patience*.

Few other hobbies and physical activities demand as much physical contact and closeness as dancing. In the classic dance pose, the leader puts their arm around the waist of the follower, and the follower places their hand lightly on the leader's shoulder. To complete the posture, they take each other's hands on the opposite side; the follower places their hand in the hand of the leader. Depending on the music, they may dance so close that they can feel each other's heartbeat. And that might make their hearts beat even faster! Dancing is also about trust. If you dance rock 'n' roll and go along for a shoulder throw, you may have a tense moment or butterflies in your stomach. If the move goes wrong, it's bound to hurt!

Partner dancing is a challenge even without throws and jumps—a play of closeness and distance. It can be an exciting experience not only for strangers dancing together for the first time, but also for lovers in a long-term relationship. Being close and gazing into your partner's eyes can do wonders. Eastern wisdom teaches us that the eyes are the window to the soul. Numerous studies have provided support for that idea. In one, human participants were able to accurately decode most emotions in facial expressions that they saw in photos on a computer screen. It turns out that pupil size differs depending on what emotion is being felt and expressed. Although the participants were not actively aware of the difference during the study, they used this information to judge which emotion they were seeing. Their brain did the work for them, noting subtle differences in the models' pupil size and using that information to decode the right emotion. Another line of research has shown that romantic partners feel more closely connected if they gaze into each other's eyes for a certain amount of time every day. Besides, neuroscience has shown that our eyes have a tremendous capacity to arouse. When did you last take a moment to lock eyes with your partner? If you can't remember, it may be time to join each other on the dance floor.

Dancing can be very erotic to watch, especially when certain moves are performed. However, studies also reveal that authenticity and genuine expression is ultimately more important than any particular move. Being yourself, it turns out, is what entices observers the most. Data shows that authentic movements seem to reveal a dancer's emotions and the truth about their physical power and mental strength. An observer's brain is able to decode this without even being aware of it. Without really knowing why, they are able to tell that they like what they see. A whole series of studies about dance and dating suggests that our movements while dancing tend to be a kind of code. If our brain decodes the information successfully, we'll discover much more about the unique "inner data" of someone's personality than we might ever find out through conversation.

A group of researchers working with Bernhard Fink at the University of Göttingen discovered that dancers' physical symmetry during a dance provides insight into their mental and physical stability. Similarly, the psychologist Nick Neave and his team of researchers were able to show that women pick up information about men's physical strength and inner qualities simply by watching them dance. The researchers first recorded a group of men dancing and then produced short video clips of these moves. Offline, the men also had to undergo a series of strength and fitness tests; for instance, they had their grip strength measured. In the second part of the experiment, the researchers showed the video clips to a group of women and asked them to rate the men's dancing. The women knew nothing about the grip-strength test, but, interestingly, they preferred the dancing of the men with the greatest grip strength. And the women in the experiment weren't the only ones who seemed able to identify the strongest man based on his dance moves. Heterosexual men's ratings revealed the same connection, which led the researchers to assume that dancing might contain coded messages to potential competitors. A study by Kristofor McCarty and his team confirmed the connection between women's preferences for male dancers and the dancers' strength (measured offline). Additional biomechanical data analyses revealed that the women's preference judgments were driven especially by men's upper-body movements while dancing.

Another surprising study by Professor Fink and his team focused on the hands of male dancers. It has long been known that the fingers of men and women have different proportions. Generally, a man's ring finger is longer than his index finger. For women, the opposite is true, or both fingers are of similar length. Researchers were able to ascertain that the difference may be due to different testosterone levels in the mother's womb during gestation. Typically, the higher the testosterone concentration in the womb, the shorter the index finger will be relative to the ring finger. Several studies now report data where female raters in experiments have ascribed a more obvious masculinity to men whose ring finger is distinctly longer than their index finger (again, the women are not aware of this difference during the experiment). A longer ring finger has been found to correlate with a higher sperm count, more aggressive behavior, more leadership qualities, and a longer penis. Women seem to be influenced by these differences when they watch men dance—without being fully aware of the reasons for their preferences. In Fink's study, women judged the dance movements of men with especially "masculine" finger proportions as more masculine, more dominant, and more attractive, without having seen the men's ring and index fingers, and without knowing about this possible connection.

Interestingly, studies reveal that how attractive a dancing woman looks to a man depends on several factors: on the symmetry of her body, on the kind of movements she performs, and on *when* during her menstrual cycle she is dancing. Fink and his team showed two hundred men video clips of dancing women (silhouettes only) that had been recorded during or after the women's fertile period. Men generally tended to find the silhouettes more attractive on clips that had been recorded during the women's fertile days. One study even revealed that erotic dancers earned more tips during the days before ovulation.

Fascinating, isn't it? We are able to read the body language of dancers intuitively and learn things about them that we'd probably never get to know during a verbal first conversation. The next time you go dancing, trust your gut: the dancer who strikes you as the most attractive might in fact be the strongest man for you. Suddenly, the Beatles' song "Something in the Way She Moves" seems to contain more truth than one might have originally thought. It turns out that a great deal of important "inner data" can be revealed through dance.

For their part, evolutionary biologists might reverse the argument: disliking something in the dance movements of a person of the opposite sex may mean that person is not the right partner for us, and we ought to trust our gut instinct.

In Argentine tango, a gentleman will ask for a dance by gazing at the lady of his choice from a distance—this is called *cabeceo* (from Spanish *cabeza*, meaning "head"; he is "heading" toward her, so to speak). If she returns his gaze, he will approach her. If she maintains eye contact, he will give an almost imperceptible nod as a greeting. This constitutes a formal invitation to dance. Be careful who you gaze at! If she wants to, the lady accepts the invitation with a nod of her own and will then follow her new partner onto the dance floor. This "invitation process" often occurs over some distance, and it is not uncommon for the partners to see each other properly for the first time only in the moment when they embrace to start the dance. That's when they become more fully aware of each other's body and personality. Once the partners embrace, something beyond the movements will determine whether a partner will be limited to this one dance or stands a chance of more dances to follow: their scent.

Cultural studies researcher Ingeborg Ebberfeld asked 432 people between the ages of fifteen and eighty-two to fill out a questionnaire. The results were surprising: whether two people have sex depends to a large degree on their scent. The chemistry has to be right, literally. How unfortunate that we currently do everything we can to mask our body odor! Body lotions, sprays, perfumes, and deodorants all serve this goal. But when we dance—really wildly and with abandon—we sweat. Sweat contains chemical signals called pheromones that can reveal important genetic information. Numerous studies have proven that fresh sweat can have a direct impact on a person's sexual arousal. Researchers were able to show the presence in sweat of the pheromone androsterone, a breakdown metabolite of testosterone. Women react to this scent with an increase in heart rate and respiratory rate. Their blood pressure rises, and their mood improves.

Sweat not only triggers arousal but also signals to potential female partners whether the man in question is Mr. Right for the dance of life. In several studies, women were invited to sniff the sweaty T-shirts of men and rank the smell from most to least attractive. The science slammer Janina Otto of the Philipps-Universität Marburg describes the results this way: "For eighty percent of women, male sweat smells of urine—and for twenty percent of very happy women, it smells of vanilla and honey."

The studies showed that the men whose sweat women perceive as attractive have genetic profiles that differ strongly from their own. In this way, nature ensures the greatest possible genetic exchange and the creation of healthy offspring.

Genetic opposites do attract. So, if you like your partner's sweat, don't be alarmed: it's a good sign. What should worry you is if you dislike it. Crazy, isn't it?

If you simply trust your senses, dancing will provide instant information about the genetic compatibility of you and your partner. It's particularly glorious, then, that most ball gowns leave a lot on display!

TO LEAD AND BE LED

Again the music makes us sway in time, and it's pretty obvious that neither of us feels like sitting much longer. Dong grins: "Do you expect me to ask you to dance just with my eyes, as in Argentine tango?"

"Absolutely not! Let's go."

We make our way to the floor for our first dance together. At first, we both seem a bit clumsy, which probably comes from both of us wanting to decide where we'll go.

"Let me lead," whispers Dong.

It works out great.

Partner dancing always plays with role allocation. Classic partner dances have a leader and a follower. In the past, those roles were assigned according to gender, and even today—especially with ballroom dance—we find separate instructions for the *gentleman* and the *lady*.

Although we no longer have a rule that dance partners must be of different genders, or around which gender has to lead, it is often the man who leads in partner dancing. To lead and to be led requires that both dancers pay close attention, because ideally the leader indicates with gentle physical signals which move, which step, or which figure will be next, and the partner makes a move in response. The follower needs to carefully observe the leader's body language to make sure their dancing will be unified and harmonious. But the leader, too, has to empathize in order to decide where the journey will go. They must engage with their partner and their partner's body language. The most important aspect of partner dancing is thus the attentive contact with one's partner. The more we dance *with* our partner, the better we get to know them.

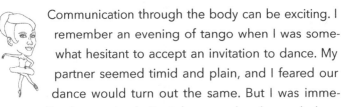

Communication through the body can be exciting. I remember an evening of tango when I was somewhat hesitant to accept an invitation to dance. My partner seemed timid and plain, and I feared our dance would turn out the same. But I was immediately surprised: Straight away, he demanded an enormous lunge, which thrust our upper bodies against each other. "I'm the boss," he seemed to tell me with that move. "Or so you think!" I thought, and responded with a playful pause. That the man is the leader in tango dancing does not mean he makes all the decisions. The man proposes the moves, and the woman may agree or interpret them in her own way. Argentines don't like the terms *leading* and *following*, because ideally the tango is two bodies engaged in a conversation. At times, their dialogue can turn into a little argument. I normally enjoy submitting to a dance partner's leadership (I have to make more than enough decisions in my work), but my partner's macho posturing made me go to the barricades. His next demand was a deep knee bend that required me to stretch one leg way back and bend my standing leg's knee between his legs. It was time to mount a counterattack! During the next bar of the music, I leaned backward slightly to clearly reject his closeness. I slowly brought my backward-stretched leg forward and placed it just far enough from his foot to force him to wait in his pose. As soon as he was able to, he responded blazingly fast with a *barrida*, a sweeping action that brought his foot very close to mine. The game ended with him blocking my leg, pulling me close, and fixing my foot in place with his hip. I stood helplessly on one leg, completely at his mercy. The music was calm, but my heart was racing... So much for timid and plain!

Men tend to find learning to dance difficult, mainly because in most partner dances they are expected to lead—and leading is the furthest thing from a beginning dancer's mind. Most men find it a big enough challenge to coordinate their legs, arms, and torso with the required sequence of steps and turns. Skilled leading only comes after one has mastered one's own steps and turns. However, someone who leads well can guide their partner to moves that the person would not have thought themselves capable of making. Good leading is so fluid and natural that the follower does not need to know the steps. In the world of tango, so-called tango marathons exist, where tango addicts meet for a weekend and dance day and night to the point of utter exhaustion, at which time they take a short nap before continuing to dance. One of Julia's tango friends recently mentioned that there seemed to be a line at those marathons that, once crossed, made dancers feel as if they were pure movement. An intoxicating sensation! Every movement seems a word, and the bodies' dialogue a

conversation. Step sequences and theories about posture, leading, and following fall away. You no longer think, he said, but you simply are, and you share that moment as equals.

In this context, swing is an interesting dance. With ballroom dances, tango, and even salsa, the movements of leader and follower complement each other; for example, the leader takes two steps forward while the follower takes two steps backward. But with swing, their steps often mirror each other: leader and follower dance the same step, but from the opposite side. Leading and following is thus possible for either partner. There is much more room for improvisation, which is the essence of this dance. If the music carries me along, I can improvise beautifully. I jump or turn and, if the dance works out, give not a moment's thought as to whether, or how, I am supposed to lead my partner. This flow is like the mirror game: you get to the point where you no longer know who is leading and who is following. Everything just happens naturally. All that's needed is a little bit of empathy for your partner.

In the end, what really matters in partner dancing is the enjoyment of moving together to music. Learning to dance takes patience and a lot of humor. In 1900, the French philosopher Henri Bergson suggested that laughter makes us feel connected, and in 2013, a group of researchers working with Dirk Wildgruber at the University of Tübingen confirmed his idea. They found that when we laugh together, the parts of our brain that are needed to

understand other people's points of view become active. This means that the laughter we share in a dance class allows us to feel closer to our dance partner. But that's not all. Laughing together also stimulates areas of our brain responsible for learning and remembering. It lets our heart beat faster and adds oxygen to our blood. All of this boosts our brain's ability to learn. It's important, though, that we laugh *with* and not *at* someone. Neuroscientific studies have shown that when someone is mocked or excluded from a group, the areas in the brain that normally make us feel physical pain and psychological sorrow are activated. So, don't laugh at your partner's awkward salsa moves; instead, laugh with each other to enjoy and celebrate your wild turns on the dance floor. And don't be afraid to laugh about your own missteps either!

> There are moments when the music makes you almost tipsy. You hear your favorite tune and you just dance. I remember one such situation. The music inspired me to try my coolest moves. I was super-psyched, felt wild and creative, and saw more eyes following me in amazement than I had ever seen before. Some folks even pointed at me. I was attracting all kinds of attention; people smiled and watched my every move. My partner, too, was in a fantastic mood, laughing along. It was an amazing feeling. I was king of the dance floor! At least I was until the end of that dance, when I noticed that I'd ripped my pants during one of my wild jumps. And right between my legs, of all places! My butt—clad in colorful underwear—was in plain sight. I still laugh out loud when I think of that evening, and of myself as king of the dance!

In a 2008 study, psychologists Eva Wunder and Klaus A. Schneewind surveyed 650 couples about the key prerequisites for a good partnership. Sharing a hobby was ranked fifth, right after tolerance, trust, love, and communication. If you are part of a couple that dances together, you experience yourself and your own body very consciously, and you get to experience your partner and their body in the same way. It's also fun to learn something new together. The choreographies of individual dances may initially seem very complicated, and it may be hard not to step on each other's toes, but things get easier with every dance. Promise! This is where the neurotransmitter dopamine comes into play. Dopamine is a substance produced by our bodies that plays a central role in learning and remembering. Numerous scientific studies have proven that when we feel a sense of achievement, our body releases extra dopamine. This in turn has a positive effect on our emotions and, most importantly, on our motivation. If you practice and learn something new with your partner, life will never get boring.

The band is playing an Argentine tango—a calm piece, nothing daring. "Do you dance tango, Dong?"

The skeptical look on his face is answer enough.

"Then it's time for me to lead!"

The tango is considered the ultimate, most prestigious of dances, but it's still all about walking with a partner, in synch with the music. And as we have seen, our brains work very well when they have to imitate a partner's movements. The mirror neurons in Dong's brain will have to start firing!

"You'll do fine!" Julia tells him. "And with a bit of luck the next dance will be a swing dance."

But really, it doesn't matter what or how we dance; the main thing is simply that we do it!

3

GROUP DANCE– THE SOCIAL BENEFITS OF DANCING

*Dance is the only art of which
we ourselves are the stuff of which it is made.*
TED SHAWN

Our academic debates today focus on the "social brain." We talk about cooperation and social interaction, and about what happens in our brains in the process. Humans are without a doubt the most social of all creatures. On our way to the bar and dance floor, we are still discussing what it is that enables us to negotiate, to compromise, and hence to collaborate.

Tonight, it's showtime. Young men and women in traditional costumes form a circle on the dance floor, ready to begin. Are these costumes authentic, or just pretty frills to entertain the tourists? The music starts. It's loud, *wild, and stirring. We immediately clap along to the two-four beat as the first dancer leads the others counterclockwise around the floor. Right from the start the group radiates an impressive energy and joy.*

DANCING FOR THE FEELING OF COMMUNITY

No one likes to be alone. Most of us know this from personal experience, but countless studies have also demonstrated how much we treasure our community. To be happy, we need other people, and we feel comfortable and secure when we belong to a group. A sense of community gives us emotional security, creates a sense of well-being, and motivates us to set new goals. In a survey by the journal *Zeit* in 2015, over 80 percent of people reported that the feeling of belonging to a community is "very important" to them. "Us" can be a couple, a family, a team, or an association of some kind. The important thing is to be together.

We are at ease when we're around likeminded people. Our brain prefers those with whom we've shared difficult, entertaining, or impressive experiences. Research shows that we find these people more pleasant, funnier, and more trustworthy, and we gravitate toward them over others in conversation.

We swing dancers regularly organize so-called flash mobs to draw attention to our style of dancing. For those who happen to pass by such a scene, the moment is magical: all of a sudden, a flash mob emerges out of nowhere. A man wearing a smart 1930s outfit gets on the subway and seemingly at random invites a female passenger to dance with him.

But she's been in the know all along. She gets up, takes off her jacket—and reveals her dance dress. Before onlookers can even start to wonder what's happening, other people arrive with washboard, guitar, and violin and begin to play gypsy jazz. At the next stop, more dancers get on, and the subway car becomes a swing-dance show.

Flash mobs give more than just the audience a terrific time. Each time we plan a surprise performance, we are as excited as children. Sometimes we've only met the other dancers online—but now we all participate in the same fun event. Afterward, we feel a strong bond with one another.

This community feeling is rooted in our evolutionary past, and is, therefore, deeply embedded in each of us. As formerly nomadic groups settled, community became more and more important. Fields were cultivated through group effort, and the harvest shared. Groups cared for the elderly and the weak, and defended their families and clans against enemies. Group bonding became necessary for survival. Rituals and traditions evolved to reinforce these bonds and to distinguish the "in" group from outsiders. All cultures, all over the world, developed group dances. Today, science shows that moving in synchrony to music or to a particular rhythm was very important for the evolution and development of human societies. Dancing strengthened the social bonds within a group and gave people a feeling of identity and belonging that distinguished their group from other clans or tribes.

If you like to watch soccer or rugby, you might know the haka, the ritual dance of New Zealand's Maori. New Zealanders still use this traditional dance to intimidate their opponents, although these days the posturing all takes place on the field.

Group Dance—the Social Benefits of Dancing

A 2017 comparative study by Yusuke Kuroda and colleagues at New Zealand's Massey University and Japan's University of Tsukuba demonstrated that the haka has definite psychological effects: after dancing haka, the participants felt more confident and dominant, full of zest for life, and pleasantly aroused.

Especially in Africa, tribal group dances are still widely practiced, and are passed on from generation to generation. Consider, for example, the adumu, a traditional dance of the Maasai warriors in Kenya. Young Maasai jump repeatedly as high as possible without letting their heels touch the ground. In a celebration that lasts for days, the onlookers accompany the dancers with songs. Whoever jumps the highest attains warrior status—and, in earlier times, perhaps the tribe's prettiest bride. The dance is a classic rite of passage that allows youngsters to become recognized as adults and accepted into the community. Such initiation rituals exist in many cultures because they strengthen communal bonds and integrate the next generation into the tribe. To this day, Western societies have debutante balls where young girls "come out" into society. Americans, especially, like such balls. One of the most impressive is the Colonial Pageant and Ball in Laredo, Texas. In a spectacle that has been celebrated since 1898, dancers wear historical costumes and follow a traditional and strict etiquette.

Group dances did not start as the abstract movement patterns we know today. Initially, they were invented to represent or re-create elements from daily life. Dancers imitated movements typical of the work done by the community or represented tribal customs through body language, and many dances tell a story. Often, the names of the dances reflect their content: potter's wheel dance, tinkerbell dance... Some of the dances were rituals or performed for healing purposes.

The development of folk dances generally went hand in hand with the development of specific traditional costumes.

Some folk dances and their costumes have become so popular that they are now recognized the world over. In the European Alps, the schuhplattler has become a tourist attraction. We associate flamenco with Spain, capoeira with Brazil, and square dancing with North America. Some folk dances, however, are not genuinely traditional but were invented more recently. The Greek sirtaki may well be the most famous example: it was first danced in the 1964 movie *Zorba the Greek* and doesn't have much in common with the more renowned traditional Greek dances, which are usually danced in a circle. As the story goes, the movie's lead actor, Anthony Quinn, supposedly couldn't cope with the complexity of those traditional folk dances, so a new dance was invented—which is now, funnily enough, widely thought of as *the* Greek folk dance. Sirtaki does have some roots in the so-called sirtos dances that are quite common on the Greek islands and, to some degree, on the Greek mainland. *Sirtos* means "to drag" in Greek and is the name of these dances. *Sirtaki* is the diminutive of *sirtos*—which fits, considering that this invented dance is a simpler form of the more complex traditional one.

A genuine folk dance, performed by people native to the culture, can be most impressive.

> It's a warm summer evening in southern Spain. The theater is dark, apart from the dimmed emergency-exit signs at the back. Suddenly a volley of steps rings out on the hollow-sounding wooden floor. I say "rings out," but it would be more precise to speak of a "shaking," because the footwork of a flamenco dancer is actually shaking the wooden platform on which we're sitting. We hold our breath, and our hearts skip a beat—even though the actual stage is more than fifteen feet away!

Another stamping assault resonates across the space. This time, the dancer's staccato catapults us into breathlessness again and again. All the while, we sit in complete darkness. Then, at long last, a light on the horizon: sunrise. Of course, it's just a stage light, but as a young woman begins to sing, her voice transforms the light into the sun. Her song is beautiful and sad, quiet at first; then it grows in power, rising into the still dawn of the scenery, simultaneously melancholic and joyous. Many in the audience probably don't understand the Spanish lyrics, but it doesn't matter. Everyone understands their meaning, each in their own way. And now a guitar player responds to the singer. Whenever she pauses, he starts up. The guitar and the voice are having a conversation—a conversation about the beauty of the moment, about the night, the dawn, and the day, about you and me. The dancer, whose stamping volleys took our breath away, now softly glides, spinning around his axis, on the sounds that fill the space and us. His arms accompany his turns with fluid movements. We no longer hear his heels. From the black void next to the stage, a delicate female dancer emerges...

Imagine a spark that appears out of nowhere, explodes into a flash, and ignites a fire. Can you visualize the scene? If so, you know what happens next. The spark was flesh and blood: "La Chispa" is a flamenco dancer on Majorca...

Flamenco was born during long nights from the stories told by nomadic peoples in the south of Europe. Inspired by the undulating arm and torso movements typical of the Arab dances of the caliphate that once existed in today's Andalusia, an art form emerged that is more than dance, music, or song:

it is all of these together, and it speaks to us about life. That is why the back room of a flamenco bar or wedding hall can become the site of an impromptu flamenco marathon. The mood is often calm in the beginning and then turns melancholy, only to ramp up into a wild frenzy. At its heart, flamenco is about authenticity and strong emotions.

Aside from its other benefits, the feelings of community that come through group dancing can motivate people to move more and, as a result, improve their health. Many African Americans, for example, consider their culture an important source of their sense of identity and belonging. With that in mind, Carolyn Murrock and her colleagues at Case Western Reserve University in Cleveland, Ohio, started an experiment in 2008 to see whether a program based on African dance would encourage African American women who did not exercise to become more active. It worked, and eighteen weeks later the women's fitness had measurably improved. In a similar experiment in 2014, Candace C. Johnson and colleagues at the University of Virginia used a culture-specific fitness program rooted in African dance. They wanted to find out whether African American women would get on board with the program and whether their health awareness would increase as a result. It did: these researchers, too, reported instances of early success.

Countless folk dances exist to this day in all parts of the world. Studying them can be a great opportunity for delving into the culture and customs of other countries. In a 2011 study by the German Sport University Cologne, high-school students were taught folk dances from all over the world. Initially, the students were highly biased and reluctant to learn about other cultures. But when they were surveyed again at the end of the program, the majority said that dance was a good opportunity to encounter other cultures and overcome bias. Dance, then, can serve as a wonderful means of integration. For example, a group of swing dancers from Berlin teaches dance in refugee camps in Greece. Dancing helps the adult refugees and, in particular, their children to experience moments of lightheartedness, to share moments of joy with people from other cultures, and to find a modicum of normalcy in their daily lives—something of which they would otherwise be deprived.

Exploring different dance traditions can help to bridge cultures. We discover commonalities and learn to respect differences. And it doesn't matter if someone is from Brazil, Japan, Russia, or Syria: dance does not require language because it is one. Dance is one of the best ways to get to know people, even if we don't speak their language.

MOVING WITH THE TIMES

The folk dance group receives enthusiastic applause and takes their bows. The group dissolves, and the dancers approach us. Every member of the group takes the hand of one audience member and pulls them on to the dance floor, to the applause and cheers of the remaining onlookers. Apparently, it's time for a Greek folk dance! The two of us have been nabbed and are grinning helplessly because right now all is utter chaos. Meanwhile, the music plays on, not missing a beat. We lock arms and copy the steps of the person next to us. Left foot, right foot, left foot, right foot. Everyone together. Shoulder to shoulder, arm in arm. Faster and faster. All of a sudden everything clicks. As if someone had flipped a switch, we suddenly move in synchrony with our Greek dance partners and hence with the group. It's not difficult to dance in time to the music because the group decides the moves. It feels fantastic and is incredibly fun.

Many folk dances are group dances, such as the salsa rueda or step dancing, as are formation dances such as square dancing or the gardetanz. Dancing with others has advantages: we are not limited to one partner and therefore not dependent on one particular person. We won't get stood up and won't stand around helplessly when our partner happens to be away; instead, we are part of a larger group. When everybody is dancing the same steps, learning is made easier for many of us; no one is forced to lead, and the dynamic energy of the dance carries us along. When square dancers stamp on the floor, the dance hall shakes. Everyone is part of the movement, and everyone's individual movements gain new meaning as we experience ourselves and our bodies in connection with others. Dancing together bonds people like almost no other activity.

Group Dance—the Social Benefits of Dancing

Three minutes of dancing can bring us closer to someone than three years of "knowing" them in everyday life. Here it is again: that feeling of togetherness.

People you have shared a passion with for years can come to feel like part of your family. The friends I have danced with have a special place in my memory, different from the friends I've never danced with. Maybe that's because we share so many emotions when we dance together. I met Nadine on stage during a dress rehearsal for the ballet *Coppélia* when we were both thirteen. We danced with each other before we exchanged a word. She was part of another training group, and we had been taught the choreography of the piece separately. Now we were dancing it together for the first time. Right from the start we liked each other and recognized ourselves in each other's movements. Today, she is like a sister to me.

Dancing together lets us experience joy, pleasure, and a distinct feeling of belonging. Can you remember any of the popular dances of the last few decades? Which one did you dance over and over? Was it "Lambada," the adaptation of a Bolivian song by the Brazilian pop group Kaomo in the 1980s? Or the Spanish "Macarena" by Los del Río in the 1990s? Or maybe PSY's "Gangnam Style" in 2012? These dances came from all over the world and got people from all over the world to move to the same beat. Without needing to exchange a word, we can get into the groove with people we've never met. Whether during a holiday on Majorca, in a club in downtown Manhattan, or in a bar somewhere in Asia, it's a powerful feeling! Dancing in a group is always an exciting experience.

In group dancing, synchrony is very important. The Greek word *syncronos* means "together in time." When we move in synchrony, we move the same way as the others and at the same time. We behave like a single, large organism. And we humans love moving together in synchrony. Whether we are marching in step, singing in a football stadium, or line dancing in a country-and-western bar—during those times when we move in synchrony, the line between "me" and the group blurs, and a strong bond is formed.

In 2007 and 2008, Jorina von Zimmermann and Daniel Richardson at University College London, as well as Guido Orgs and his colleagues at Goldsmiths, University of London, published fascinating research on dance and synchrony. They had people who did not know each other dance freestyle while wearing measuring devices on their wrists that recorded how synchronized the individual dancers' movements were with each other. The more synchronous the dance movements, the more the dancers liked each other and the more easily they agreed with each other in the exchange of ideas after the dance session. Generally, the dancers who had the most synchrony with the movements of the group overall also felt the strongest sense of belonging to the group. People who dance together become more social, have more positive feelings toward others in their group, and even develop similar views.

In 2014, evolutionary biologist Bronwyn Tarr and her colleagues at Oxford University were able to explain how it is that dancing leads to the development of such strong feelings of community, even if none existed before. When we are simultaneously aware of our own movements as well as identical movements performed by another person, so-called coactivation occurs in our brain: areas are simultaneously activated that pertain to awareness of "self" and to awareness of "other," though these areas usually operate separately from each other. The effect of this coactivation is a blurring of the line between our awareness of self and our awareness of others. Put differently, when we dance together in synchrony, our brain perceives self and other as a single unit. Studies conducted by evolutionary psychologist Robin Dunbar and his team—as well as other research teams—have shown that this effect occurs only when we move in synchrony; it does not occur when we move asynchronously, or are out of synch. Synchronous dancing lets us literally *merge* with each other. When we dance in a group, we will eventually feel one with the group... we no longer perceive ourselves as separate individuals but as part of the larger whole: "we" instead of "I."

Psychologist Scott Wiltermuth and his colleagues at Stanford University were able to demonstrate in a variety of cooperation tests that synchrony rituals support the formation of stable groups and communities. The researchers asked a number of their research participants to engage in shared activities. Some danced and sang together, while others marched in step. Afterward, the researchers tested how well the participants were able to act collectively in a number of different situations. It turned out that members of synchronously acting groups showed greater loyalty toward each other than participants who acted individually. The synchronous group movements also prevented people from stepping out of line and enriching themselves at the expense of others. This, again, is the mirror neurons at play: synchronization is, after all, mutual imitation that demands everyone's empathy and promotes mutual understanding. Jonathan Haidt of the University of Virginia even hypothesizes that coordinated behaviors such as group dancing and singing are a special type of pleasure that we only get through synchronized action. In 2009, Vasily Klucharev of the Radboud University Nijmegen was even able to show that conformity (a type of synchronized action) activates the brain's reward systems.

In 2017, psychologist Jan Stupacher studied whether music can reinforce these social bonds created by synchronized movements. His results show that the most intense social bonds are created when people move in time to music. And a 2012 study by Alessandro D'Ausillio and colleagues showed that listeners perceived orchestral music as more beautiful and harmonious if the musicians moved in synch with the conductor's movements while playing their instruments. Interestingly, this synchrony even affected the listeners when they weren't able to see the movements.

TEAM SPIRIT

What researchers are now learning through scientific methods has been known to dancers for a long time: dancing together creates strong bonds. Interestingly, observational studies show that in situations where group cohesion and team spirit are required, people often spontaneously choose to dance—in kindergarten classrooms or seniors homes alike. Also, for groups that struggle with social challenges and conflict, the right dance project can cause small miracles.

And dance can do even more. Bronwyn Tarr and her colleagues at Oxford University examined the effects of different degrees of physical demand during synchronous dancing. They divided 264 Brazilian university students into groups of three and assigned them to one of four different categories. The students in category one were asked to dance in synchrony with a high level of physical demand; in category two, they were asked to dance synchronously while seated; in category three, to dance asynchronously with a high level of physical demand; and, in category four, to dance asynchronously while seated. The students were then asked how close they felt to

Group Dance—the Social Benefits of Dancing

their group members, and blood-pressure cuffs were inflated to measure individual pain thresholds. Exerting a high physical effort while dancing in synchrony noticeably increased participants' pain threshold. The researchers interpret this as evidence for increased endorphin levels. Endorphins are morphines produced by the body that act as pain relievers or even pain suppressors. They regulate hunger, help modulate our mood, and contribute to the production of sex hormones. Endorphins are produced in the brain and are also activated in emergencies when we enter a state of shock. They are one of the reasons that, immediately after an accident, people may not notice how badly injured they are. Endorphins are also released as a result of positive experiences or exercise, which is why they are—misleadingly—called happiness hormones.

Based on extensive research, researchers assume that synchronous movements, too, cause the release of endorphins. This, in turn, conditions us to develop a positive attitude toward the people with whom we've danced. This means that synchronous group dancing makes us more social.

"Dancing has direct effects on our brain," says Julia, summarizing the insights of the day. Watching the effects of group dancing is fascinating. Even though we can't measure endorphins here on the dance floor, or prove that team members magically bond together, it is clear for all to see: everyone is in a great mood after the folk dance interlude, giggling like goofy teenagers. We'd all been dead tired after a full day of presentations, but from the speakers now comes a familiar tune, with some very familiar lyrics about a young man who has no need to feel down. Almost all of our colleagues are already on the dance

floor, and if we want to further the team spirit, it's time for us to join them. By the time the refrain comes around, we have, and now we are laughing in the midst of a dancing crowd where everyone is waving their arms in synchrony to "YMCA."

4

DANCING FOR THE BODY–DANCE AS AN ELIXIR OF LIFE

· · · · · · · · · · · · ·

[When I dance], I sort of disappear,
I feel a change in my whole body... there's fire in my body...
I'm just there... I'm flying like a bird, like electricity.
FROM THE FILM *BILLY ELLIOT*

· · · · · · · · · · · · ·

It was quite a late night–or rather, an early morning! Back in the seminar room, we're sleepily squinting as the sun shines through the windows, and having a bit of trouble concentrating.

"Everybody up, please!" The neuroscience professor who is set to deliver today's introductory lecture leads us through a stretching exercise before she starts.

"I know, I know. You're perfectly comfortable sitting there," she concedes. "It'll be quick. Come on! Arms above your head, please."

She demonstrates what to do.

"Now stand tall and stretch sideways. Well done! Now look up at the ceiling, and remember to stretch your neck, too."

Knees and elbows creak. Our sore calf muscles feel the aftereffects of the Greek dance interlude from the day before. But our bodies loosen and open up. There's laughter, and faces grow more alert. Then we are allowed to sit down, and the professor begins her lecture.

I LIKE TO MOVE IT, MOVE IT!

No one questions that movement somehow feels good. We all know that staring at a screen for hours on end isn't exactly great for our health. We feel guilty if we spend a whole weekend on the couch watching TV or in the sun on a deck chair reading a good book; after all, we read and hear all the time that we ought to move. To exercise. To become more active. But why?

As a former dancer, our neuroscience professor knows very well that being physically active works wonders for body and soul; dancing is, above all, movement, and hence an effective workout for the whole body. Dancing improves our fitness, flexibility, and coordination. On the dance floor we can work up a sweat, and such regular physical activity builds stamina. Fortunately, it's also easy to determine how hard we want to push ourselves. We can be relaxed dancing the blues or go full out with a fast jive. Competitive dancing means high-intensity performance training. Movement makes our heart more efficient and with time improves circulation, which lowers blood pressure more effectively than medication does. The systolic pressure (the top number in a blood pressure reading) can be lowered by up to 10 to 15 mm Hg, and the diastolic pressure (the bottom number) by 5 to 8 mm Hg. Not only does our heart pump rapidly when we abandon ourselves to the beat on the dance floor, but we also breathe markedly faster, which in turn increases the body's supply of oxygen.

When we move, a lot happens in our bodies. We may be unaware of some of it at the time, but it nevertheless benefits us in the long term. For a start, dancing makes our leg muscles work hard, as anyone who has danced a night away can testify. Although we "only" danced, we may feel sore the next day. But dancing involves many more muscles than we may realize in the moment. Whether enjoying ourselves on the dance floor of a club, practicing ballet at the barre, or dancing a rumba with a partner, certain postures are required: we have to tense the muscles in our buttocks, abdomen, and upper body or it won't work. Dancing is thus belly, butt, and leg training, and in the process strengthens our back.

My ballet teacher used to offer me this mental image: "Julia, imagine me pulling you toward the ceiling by your hair." As a child I found that easy to grasp, and even today, when I'm sometimes slouched at my desk, my ballet teacher appears in my mind and instantly makes me straighten my back, shoulders, and neck.

Anyone who dances regularly improves their posture. Dancing has us stand tall and makes us perform various movements simultaneously that require coordination. We spin around, stand on one leg, or lean into our partner's arms. All of this improves our sense of balance and perception of our own body. Researchers now know that muscles are not merely "power packs" that allow us to move with precision but that they do much more for our bodies. A team of researchers from

Copenhagen discovered that during physical activity skeletal muscles produce certain messenger substances, so-called myokines. This field of research is still in its infancy, and we are not yet exactly sure how these myokines work. The myokine we understand best is interleukin-6: it strengthens the immune system and is anti-inflammatory. It also affects our sugar metabolism by making muscle cells absorb glucose from our blood. This way, the myokine can protect us from developing diabetes. Speaking of diabetes: in 2014, Felice Mangeri and his colleagues from Brescia Hospital and the University of Bologna demonstrated with their BALLANDO program that two hours of Latin dance training twice a week had important health benefits for diabetics. Not only did the study participants show improvements in their fitness data and a reduction of their disease-related problems, but these improvements were still present after three and even six months, something that is not the case with conventional diabetic fitness programs.

Regular dancing strengthens muscle cells in the entire body and trains our muscles by having them interact in natural movements rather than providing the narrow, partial muscle training of fitness machines. It gives muscles more power and lets them react better and faster when stressed, which also enhances coordination. Our muscles and organs are all surrounded by fascia—a delicate white "skin" that functions as conjunctive tissue. Fascial tissues tend to glue together as we age. And because more nerves pass through the fasciae than through just about any other kind of tissue, our fasciae are full of pain receptors. The fast movements in dance offer particularly good training for keeping our fasciae supple. The blood supply to tendons and ligaments, too, improves with rhythmical movement, which leads to much better flexibility. This, in turn, protects our joints. You probably all know a jogger who has ruined their knee or a tennis player who wrecked their elbow. Those kinds of injuries are less common with dancing. Some sports scientists therefore see dance as distinctly more beneficial than other types of sport. When we dance, we leap and spin around, raise our arms, bend our knees, and flex our back. What other kind of sport has such a huge range of motion? As a result, dancers generally avoid unilateral stresses.

A VERY SPECIAL NERVE

When doctors in the Middle Ages began, in secret, to dissect corpses to learn more about the body and its functions, they kept coming across one nerve in particular. Today we know it as the vagus nerve. It is the tenth of twelve cranial nerves, and its name—derived from the Latin *vagari*, "to wander"— describes its purpose: it wanders (or "vagabonds") about inside our body and has countless branches. Originating in the brain stem, it runs along our spinal cord and reaches into most of our organs. Inside our head it connects with our voice box (which enables us to speak) and with the nerves and muscles in our face (responsible for our facial expressions). From the spinal cord, the vagus nerve branches out into our heart and lungs, the spleen (crucial to our immune system), and runs all the way down to our digestive tract. It connects to many of the large muscles of our body, which are responsible for posture and movement. Because it plays an important part in controlling our organs and allows for their relaxed functionality, the vagus nerve is sometimes referred to as the relaxation nerve. For many medical researchers, including the American Stephen Porges, the vagus nerve is a minor miracle; they see it as influencing the efficiency of the regulation of blood sugar, decreasing the risk of stroke and heart problems, lowering blood pressure, and promoting healthy digestion. Its activity positively impacts our gut biome, reduces inflammation, improves our mood, and hence mitigates stress.

So how can we activate the vagus nerve? A vast number of stretching exercises for this purpose can be found online, and exercise programs from the Far East, such as yoga, identify poses that stimulate our biological systems and in particular our vagus nerve.

And as you may have already guessed, dancing, too, works wonders. For example, when we bend backward or forward, we stimulate our digestion because important nerve connections reach from the spinal cord in our lumbar region into our digestive tract. When we lean far back, we stimulate the bundle of nerves below our sternum, which in turn sends up a "wake up" signal to all organs. Our heart gets stimulated and our circulation improves. Bending backward also makes us take deeper breaths. As a result, fresh oxygen from the small air sacs in our lungs enters our blood and is carried to the cells, where it is needed for our metabolism. The carbon dioxide produced by the cells is brought back to the lungs and released when we exhale. In addition, our blood provides our cells with nutrients, transports messenger substances and hormones, and removes waste products from the cells so they can be filtered by our kidneys and excreted from our gut. This process speeds up when we dance because, by moving, we increase our heart rate and breathing. Compared to doing straight muscle training and many other kinds of sport, dancing often involves raising our arms above our heads. When we do this, our lungs open, and when we next inhale, air enters deep into the lobes of our lungs. This, too, makes breathing more efficient. Deep breaths all the way down into the stomach area relax our diaphragm, the so-called breathing muscle located between belly and chest. If we hold tension in our diaphragm, we also tend to feel ill at ease emotionally. We figure something is wrong with us;

we feel trapped or constricted. Go dancing and take deep breaths! It can feel like a genuine release.

All of these movements wonderfully stimulate our vagus nerve, which has a calming and relaxing effect on our body.

Interestingly, we find the backward bend and the raised arms in images of dance all over the world, regardless of culture, era, or specific dance form. Humans seem to like these particular movements—maybe because of a positive effect on our health that we may sense, even if we are not consciously aware. Why not give it a try?

Sit on a chair and place your hands on the edges of the seat, next to your thighs. Now imagine inhaling your favorite fragrance. Enjoy that scent for two or three more breaths. Continue to breathe until your shoulders relax. Then lightly push down on the chair, relaxing your shoulders so they move even farther away from your ears. With your next inhalation, imagine a beautiful, bright light or a gorgeous necklace resting against your chest. You want to show this light or piece of jewelry to everyone, and so you move your chest as if pointing the light at the ceiling, or as if your sternum were a flower bursting into bloom. Keep breathing. Grow out of your chest and support the movement with your arms by lifting them high above your head like the branches of a tree. Don't be concerned if the unfamiliar movement and the deep breaths make you a little dizzy. Just keep breathing down into your belly. Then slowly return to your starting position.

Well done! You have just performed a ballet move—and probably relaxed your diaphragm beautifully in the process. And there is another fantastic side effect: if you raise your eyes to look up during the backward bend, you also stimulate your optic nerve.

Another movement we find in many dances around the world is the opening of the pelvis. This is clearly an erotic signal—as stretching in this way emphasizes the genitalia. But beyond this erotic component the movement has a restorative function: circulation improves in both the pelvic floor and the genitals, and we train the flexibility of our largest joint second only to the knee: our hips.

Dancing also involves movements we rarely perform in everyday life. Or do you regularly walk backward at home? Walking backward is part of the movement repertoire of dance styles such as ballroom and ballet. And that is a good thing, because when we sit—as most of us do most of the time—our ilia, or big hip bones, are pushed forward. That puts pressure on our spine and eventually causes back pain. If you walk backward, you reverse this effect. The big hip bones return to their natural position when we stretch our legs backward, and that takes the load off our backs. Walking backward has other great side effects: a study by Wei-ya Hao and Yan Chen of the Institute of Sport Science in Beijing showed that after eight weeks of walking backward every day, seven-year-old children had markedly improved their sense of balance. This benefit was still in place three months after the daily practice had been discontinued.

In addition, Davide Viggiano and his colleagues at the University of Molise in Italy were able to demonstrate that youths with attention-deficit hyperactivity disorder significantly improved their ability to stay focused and reduced their impulsivity, as measured in objective tests, after a two-month training program that included ten-minute backward walks.

So backward walking obviously trains your attention, another free benefit of dancing.

> I love to dance tango. One dance can last up to fifteen minutes, and in an evening of tango I will enjoy on average six to ten dances. Being the woman, I will move backward most of the time—and now I understand why my focus has definitely improved since I started tango dancing!

So, starting today, go for one thirty-second backward walk daily, from the bathroom to your desk... or else, just go and dance, and the backward walk will be included free of charge.

MUSIC IN OUR HEAD

> The conference sessions today are exhausting but hugely interesting. It's fascinating to learn about what happens in our brains when we deal with stimuli from our environment and relate them to previous experiences. "YMCA" accompanies us through the day as people whistle, hum, and sing the 1970s earworm. After dinner, we head almost as a matter of course to the bar and continue to chat about music. Music is more than a hobby for the two of us. We can't imagine our daily lives without it, and we enjoy it just about everywhere and all the time: when we exercise, relax, even cook. And, naturally, when we dance, for most people cannot imagine dancing without music.

Music carries us away, triggers floods of memories and emotions. It directly affects our brain. Listening to music activates different brain regions with widely diverse functions and triggers associations and emotions. When we listen to a sad song that gives us goosebumps and almost makes us cry, different neural networks are involved than when a throbbing beat tempts us to dance.

How music reaches the brain, which brain regions it activates, and what effects it has subsequently on the body is highly complex and still being researched.

Thanks to the use of early imaging techniques, researchers discovered that the brains of musicians look different from the brains of non-musicians. Most likely, people are not born with such different brain structures (and thus lean toward becoming a musician as a result), but rather it is the music that over time effects changes in the brain. Today we believe that making music, or even just listening to music, is one of the most powerful ways to change the structure of one's brain. When we listen attentively, our brain creates new neural connections, which initiates learning processes. These enable us, for example, to categorize a piece of music after just the first few bars. Anyone watching a formal dance can see how this works. When the band begins a piece, it takes the dancing couples only seconds to know, "Oh, four-four time. Latin. Judging by tempo and stress, a cha-cha." And off they go! In that moment, we don't even realize what a little miracle it is. We recognize in a flash something highly complex because we have such a huge library of music in our brain (any iTunes library would fade into the shadows in comparison). Our digital collection might include Beethoven and U2 and Ariana Grande, but our brain's collection includes so much more: that song we could barely make out above the chatter at a café in

Dancing for the Body–Dance as an Elixir of Life

Paris, the record the DJ played at our best friend's wedding, and the folk tune that our grandmother once hummed while rocking us to sleep. Every one of our musical experiences—all the musical patterns and rhythms we have ever heard—is stored in our brain. And we expand this library continuously throughout our lives. When the band starts playing, we don't need to know the performer or even the song—within seconds we draw on the musical library in our head for comparison and start to dance.

Music is an enormous stimulation for our brain. Countless studies have proven that it impacts our hormonal balance and decreases stress more effectively than any medication. Music also enhances our power of concentration and our emotional and linguistic competence. Children who play musical instruments find it easier to learn another language. What's more, we reap the benefits of music not only when we play a musical instrument or listen to music with full attention, but also when music is merely on in the background.

Above all, music makes us happy because it offers us a whole cocktail of the so-called happiness hormones. It puts us into a good mood and lets us forget that we are actually working hard when we dance.

> Whenever I hear piano music, I remember my first ballet classes. I am five years old and opening the large wooden door to the ballet studio, smelling the familiar scents of leather slippers and cleaning agent. I sense the soft cotton of my beloved leotard that clings to my body with unfamiliar tightness. I am back in the studio in Flensburg, a room that despite its large windows was always cast into semi-darkness and whose large barres with their black cast-iron feet were far taller than me. How much I loved that room! Any

time I close my eyes and listen to a piano—and when the music has a bit of an echo, as it did in that almost empty hall—I am transported back to that place. As if I had time-traveled!

Music is not the only thing that makes dancing so beneficial. Dancing is also a great hobby because we typically are not alone when we dance. Whether we join a folk-dance group, take dance classes, or go clubbing on Saturday nights, we meet other people, make new acquaintances, and laugh together. For people who lead rather private lives, especially, dancing can be a social highlight. Add to that the fact that we often touch each other when we dance. These physical touches are not accidental, like at work or on public transport, but deliberate and often very mindful. Our dance partner takes our hand to lead us to the floor. One partner lays a hand on the other's shoulder blade, and the warmth of that hand is felt through their shirt. A quick caress, a firm hug, or an appreciative pat on the back may acknowledge a particularly successful leap. We have body contact when we dance, and we move in synchrony. Anyone who believes that such physical touches are mere luxury is mistaken. Apparently, a human being needs eight hugs a day to stay healthy. Loving physical contact triggers the release of various vitally important biological substances. These substances enhance our social bonds and our feelings of well-being and strengthen our immune system.

I am writing these lines in my favorite coffee shop. The place is bustling. A little girl snuggles up against her dad. She tugs at his arm to make him hug her. When he gives her a big bear hug, she visibly relaxes. While her dad continues to talk with his friend, the girl looks at the world from the safety of his embrace. She watches the waiters rush about, grins at a grandma's big hat, and gazes with wide-open eyes when a cake is carried past. Occasionally, her dad tenderly strokes her hair. It looks as if she barely registers his touch, but in reality, these moments create some of the strongest biological effects that can occur in the human body. They may well be part of what makes us human. Even more importantly, we cannot survive without physical contact with other human beings.

In the 1950s, the American psychologist Harry Harlow tried to get to the bottom of the importance of physical touch. He placed young rhesus monkeys into cages without their mothers but with two different dummies: a substitute "mother" fabricated from wire who held a bottle with milk, and another substitute "mother" of equal size whose wire frame was covered with fabric but who offered no milk. The monkey babies went to the wire frame solely to drink but otherwise preferred to stay with the cloth dummy. The most distressing thing about the experiment was the fact that the motherless monkeys not only became fearful, hysterical, and later apathetic, but they also soon died. First to die were those who only had access to the wire "mother," while those with the cloth "mother" perished later. When their bodies were examined, they were found to have gravely weakened immune systems. They had died from common diseases and infections.

Since Harlow's experiments, a series of other studies have confirmed that touching and stroking appear to have important biological functions for mammals. In a 2007 experiment, French social psychologist Nicolas Guéguen had his assistant ask women in a nightclub for a dance. The women were more likely to accept if the assistant touched their arm lightly when extending the invitation. This strategy convinced 65 percent of the women, while only 43 percent of those who were not touched on the arm agreed. While this experiment clearly took place before the #MeToo movement, it underlines the importance of touch. Physical contact can get "under our skin" because it triggers biochemical reactions in our brain—the release of messenger substances that make us feel good. We feel affiliated or connected to the one we touch, or to the one who touches us. Dopamine and the bonding hormone oxytocin are just two of these messenger substances. Our sense of touch is the first sense we develop in the womb. Only later do we develop our senses of taste, hearing, sight, and smell. Human beings need to be touched. Without touch, our bodies and souls suffer. Dance offers a wonderful opportunity to touch each other.

BODY PERCEPTION

After a whole day in an air-conditioned lecture theater we feel as if we're wrapped in cotton. Most of the time the room is shielded from natural light, and the outside world seems muted. Today's lectures dealt very much with perception—that is, with our senses and what we see, hear, smell, taste, and touch. Of course, we perceive much more. Here on the dance floor, music carries us into another world. Our hearts beat, and the sense of pleasure thrills and gradually spreads inside us until it takes over.

Movement, music, and physical touch—they all affect us in important and different ways. What makes dancing so special is the active involvement of each of our senses. Our ears listen to the music; our eyes gaze into our partner's eyes, look at the band, or follow what is happening on the dance floor; our nose takes in our partner's scent or the slightly chemical smell of the dry-ice fog. As our partner leads us, we feel their hands. We swirl through the room. Maybe our mouth moves because we sing or kiss or whisper... Our senses are heightened; we pay attention and are completely in the moment.

Dancing piggybacks on psychological and biochemical processes that are already unfolding within us and can regulate or stimulate them. While we blissfully give ourselves over to the moment's groove, small miracles take place in our bodies.

In response to sensory input, our brain sends messenger substances to the rest of our body: cells are to be repaired, stress hormones to be reduced. Movement and music trigger the production of endorphins that make us relaxed and content. This experience is sometimes called the "runner's high" because the trancelike state is well known to long-distance runners. It comes with a release of dopamine, which increases our motivation, and of serotonin, which revs up our feeling of physical well-being. Together, these substances give us a "flow" experience: everything works like a charm. We feel happy.

In his 1990 book *Flow*, Hungarian American researcher Mihaly Csikszentmihalyi describes how philosophers, scientists, and artists have often claimed that moments of "flow" can bring us real joy and physical well-being.

Do you remember the dance film *Billy Elliot*? In one scene, the strict examiner at an audition asks Billy what it feels like when he dances. "I forget everything. It all disappears, and I feel a change in my whole body... there's fire in my body... it's there... I'm flying like a bird, like electricity." It would be hard to come up with a better description of "flow."

Generally, we can sense processes in our bodies. We can feel our heartbeat or notice our stomach growling when we are hungry. We sense when we sweat or begin to blush. The ability to register our body's signals is called interoception and is crucial for our sense of well-being. Interestingly, this sensitivity varies somewhat from person to person because interoception is shaped by our individual history and learning experiences. This means that we can deliberately influence it to enhance our awareness. Again, dancing comes into play, for several studies were able to show that dancing adds to an improved awareness of our body and our sense of self.

Professors such as Hugo Critchley, Bud Craig, Manos Tsakiris, and their teams have discovered that our awareness of our bodies' biological processes is connected to our awareness of our emotions. People with better interoception are more empathic and sensitive toward others, more likely to help people in need, and generally appear more capable of coping with life.

These research findings highlight how important it is to be aware of one's body and to develop a sensibility for its processes.

In one of our studies we were able to find that dancers have better interoception than non-dancers, or have a "sense of emotion," so to speak. "Dancers have a sixth sense," ran the headline of an article about the study. The existence of such a sense may be hard to accept when we think of alpha males charging around the dance floor of a club without any consideration for others. But as we observe ourselves dancing, we learn more about our body and its needs and deepen our awareness of our feelings and motivations. Do we use wide, expansive gestures or small, timid ones? Do we dance with solid steps, stomping our feet, or do we glide across the floor? How do we feel in the process? Sometimes it is only when we dance that we notice how stiff and tense we are at that moment. Dancing can make us more aware of what lies close to our heart or weighs heavily on our mind because in dancing we act out emotions and hence train our awareness of them. Dance is therefore also considered a form of catharsis.

Beyond that, human beings have a particular sense of perception that tells the brain how the body moves in space. Thanks to this sense, movements leave "impressions" on the brain's motor areas. This type of perception also improves with dancing, and in fact seems to serve as an additional sense for dancers.

To further explore this sense, neuroscientist Corinne Jola at University College London conducted an experiment in 2008 that involved both professional dancers and non-dancers. The participants sat at a table whose top was marked with five dots. They were asked to keep one hand under the table, and with the index finger of that hand they were to tap as precisely as possible the location of one of the dots. The researcher varied the conditions: In the first round of the experiment, participants were blindfolded while the researcher moved the index finger of the participant's other hand on top of the dot so that the participants had to rely entirely on their interoception. In the second round, participants were able to see the target dot; and in the third, they were able to see and in addition were allowed to use their second index finger to help. The results were surprising. Professional dancers did better than non-dancers only if they were blindfolded. When they were able to see the dot, the dancers still relied on their body perception and, as a result, did more poorly than the non-dancers, who used both their visual sense and their interoception. For the dancers, interoception appeared to be even more important than their sense of sight. The good news for all weekend dancers is that even amateurs can improve their interoception by getting out on the floor.

DANCING MAKES US SMART

The music is making Dong melancholy. So far from his family, he is missing them, and he talks about his son, Theo, and shows us photos on his cellphone.

Everyone agrees that Theo is very sweet, but one of the snapshots is particularly interesting for us neuroscientists, because we can almost see what is going on in the little guy's mind. The photo shows Theo sitting in his highchair in the garden. His mother is arranging a bouquet of flowers, and the boy seems eager to help. Additional roses are scattered on the table and on the ground around Theo. He holds a rose in his hand and offers it to his mom, and his face is lit up triumphantly: I've got it!

As we saw in chapter 2, movements are a means to explore and eventually comprehend the world around us. The momentary feeling of delight when we accomplish something signals to the brain that the movement we just completed was done "right." For Theo, a very precise amount of reaching and grasping were necessary in order for him to compensate for his position in the highchair, access the rose, which was about ten inches away, and take it in his hand. That successful sequence gets stored in the memory system's neural wiring, which is connected to the brain's motor systems. Grabbing the next rose will be that much easier, require less concentration, and succeed at the first attempt. The brain learns through trial and error. No climbing of stairs without stumbling first. No socially acceptable eating with a spoon without Nutella all over one's face first. No ball in the goal without balls missing the goal first.

The brain of a newborn consists of 100 billion to 150 billion nerve cells, the neurons. Initially, these cells are not wired together in any stable form. They are a bit like an IKEA shelf we haven't yet assembled: all the elements are there, but the screws are still in their bag, the boards piled on top of each other, the posts lined up back to back. Everything is close but has yet to be put into its proper position to allow for its later function.

Learning is at first nothing but the brain's reaction to experiences. Experiences enable the formation of new connections between the nerve cells. A single nerve cell may be linked to hundreds of others. This is how complex neural networks are formed in our brain. We can visualize the process of network formation as a lake that freezes over. First there are individual pieces of ice. Then, slowly, larger ice floes emerge, linked by slight tendrils. Finally, strong and stable connections form, allowing us to walk and quickly glide from one place to another. It is similar with neurons. Every time we learn something, neural connections are created. But in contrast to ice, whose crystals are rather hard and inflexible, neural networks are extremely flexible. They can form ever new connections.

The more often we have the same experience—the more often we reach for the rose—the more stable the neural connections will become. Researchers were also able to show that the more senses are involved in the learning process, the better the newly learned material will be stored. Any form of movement leads to neural networking. Especially in the first years of life, the development of children's motor skills is closely connected to the development of their cognitive skills. Children with more opportunities to move and to explore their environment through movement will later have better memorization skills, be faster and more creative in their problem-solving, and be more capable of concentrating deeply and for longer periods. The same is true for adults: the more we move—that is, the more we have experiences involving movements and observe them—the more our cognitive skills improve. But can we still develop new neural networks in adulthood?

When I was at university, I no longer wanted to have anything to do with dance. Ever since my accident, reminders of dancing had caused me great emotional and physical pain. Even listening to classical music was painful. But I became more and more discontent because my body needed exercise; of that, at least, I was sure. I could feel my creativity vanishing, and my thought processes become sluggish. I sensed a big difference between my current state and the ease with which I'd previously thought and problem-solved. Without dance training, my thoughts had become as flabby as my belly. But what kind of exercise should I choose? I tried swimming but found it too wet; jogging felt too one-sided; and weight training at the gym struck me as dull.

One day my Andalusian fellow student and friend Cristina suggested a new training program offered by the university. "They offer Zumba now," she said, waving a flyer in front of my nose. After the first lesson we both lay on the floor (sitting was no longer an option), sweaty and red in the face and laughing—but I was deeply grateful to Cristina. This was no ballet, no sophisticated art of movement, but it was dance! We had moved to music—sweat and happiness included! After a few weeks I noticed that regular Zumba dancing had improved my concentration and that those flashes of inspiration and lightbulb moments had returned. I was thrilled!

Zumba is a dance exercise program for fitness that was created in the 1990s by a Colombian dancer and choreographer. Thoroughly athletic, Zumba combines aerobics, Latin dance moves, and conventional fitness exercises with the rousing sounds of Latin summer hits. In Zumba dancing, our ears listen to wild rhythms, and our eyes follow the instructor's movements as we try to mirror them. Given the tempo, it is challenging to concentrate on the correct step sequence, not get one's arms and legs into a tangle, and not lose sight of the big picture. The brain must activate its perception of time and regularity to recognize the rhythm, while we must keep an eye on, and wordlessly communicate with, our fellow participants so as not to collide with them. So many skills are required simultaneously: coordination, intuition, logic, sense of time, spatial awareness, musicality... the list goes on and on.

Combined, it all leads to new neural networks in the brain that demonstrably change its physical structure. This applies to both children and adults. Our brain's ability to adapt to new challenges remains in place throughout our lives. It is called neuroplasticity. Whereas people used to believe that brain cells die with age, we now know that neuroplasticity enables us to learn new things well into late old age.

The German medical doctor, moderator, and bestselling author Dr. Eckart von Hirschhausen had the courage to do an experiment on himself—and share it with the world via his television show. The doctor was curious about whether dancing would really change his brain. To measure his brain activity, he agreed to put himself into an MRI machine on two separate occasions. Between the first and second, he took dancing lessons and learned a dance he had not known before: West Coast swing. He had not set foot in a dance school in more than thirty years. "At first, I kept wanting to check where my feet were. But luckily," he said, describing the challenge, "my feet have a direct connection to my brain, so pretty soon I was able to step on my dance teacher's toes with my head held high." Then everything clicked into place, and Eckart von Hirschhausen danced as if he'd never done anything else. What was really exciting, though, were his MRI scans. Before his dance lessons, only his brain's acoustic and visual regions were active when he listened to the music and imagined the way the dance would look. The scan after his lessons registered a big change. "I no longer imagined the dance as an uninvolved spectator; after the training, my brain translated the music automatically into patterns of movement. No longer an observer, I was right in the thick of it; instead of watching, I was twitching and lighting up all the way from head to toe." His brain's pyramidal tract, he explained, showed fine signs of a thickening of the wiring from his motor cortex to his muscles. The dancing had clearly made a difference. With his before-and-after experiment, Eckart von Hirschhausen had demonstrated to his audience the neuroplasticity of his brain.

If we want to stay mentally fit, it pays to challenge our brains. Even dancing as a hobby can bring important benefits for our cognition and the development of our intelligence because it makes comprehensive demands on our brain. The brain receives different sensory inputs, and motor processes are reinforced.

Intelligence means, among other things, the ability to come up with creative solutions to problems. That requires quick decisions and great cognitive flexibility. When we dance, each moment demands a decision. Where shall we put our weight? Where our leg, our arms? What's the next step? Such decision-making is good support for our brain.

<u>Interval training, or regularly alternating periods of load and</u> rest, is particularly effective in stimulating our neuroplasticity.

Dance is always a kind of interval training because we never do the same thing continuously; instead, one moment we skip, the next we run, we walk, or we may just stand still during a pause in the music. But another advantage of dance is that we repeat certain movements. Already in the 1950s, cognitive scientist Donald Hebb noticed that nerve cells that are activated simultaneously will form stronger connections, or, as he put it, "What fires together wires together." To apply this observation to dancing, the more we practice the two-step or the reverse turn in the waltz, the more easily we will remember the step sequence and learn the movements.

In 2013, neuroscientists working with Hubert Dinse at Ruhr-Universität Bochum were able to show that dancing improves not only our fitness but also our attention and responsiveness—even with just one hour of training per week! Other studies by a number of researchers have ascertained in the last few years that dancing improves memory and even increases creativity. In sum, dancing not only makes us better at dancing but benefits us in a whole range of areas.

But learning to dance takes time. Strong neural connections don't grow overnight. We learn by doing something repeatedly. In the beginning, we mostly learn by imitating others. For children all over the world, imitation is the order of the day. External stimuli will make a child imitate what he or she perceives. If Dad smiles, the baby will sooner or later smile in return. If Grandma waves, the baby will wave back. If the big brother skips, the toddler will try to skip as well and, after perhaps a few tumbles, eventually skip in competition. From an early age, we imitate.

Even when we learn abstract things, such as mathematics, we follow other people's line of thought until we "get it." The same holds true here: "What fires together wires together." If our neurons fire together often enough, they form a network, and eventually we can solve new algebra problems by ourselves.

Tango dancer and psychologist Nicola Clayton and her colleagues at the University of Cambridge describe our ability to imitate as the cognitive foundation for our ability to dance. In order to dance, we need to take what we see and what we hear and combine it with movement—that is, we need to be able to imitate with our body both another person and a rhythm. We need to be able to synchronize and harmonize our own movements with those of others. This ability is pretty much unique in the animal kingdom. The species that comes closest to having this ability is birds. A study done at Harvard University analyzed more than five thousand YouTube videos to find out which species is in fact capable of performing synchronized movements to various pieces of music. With the exception of one young elephant, it was exclusively birds who were able to "dance." The researchers suspect that this is because birds are the only species besides humans capable of imitating sounds.

Other studies were able to uncover a direct connection between the ability to dance and the ability to imitate, as well as language skills and social behavior. The more an animal species is capable of imitating behaviors or even sounds, the more complex its communication skills and social behavior—and the more numerous its dance-like movements. However, only humans have the ability to express something through dance. Birds can imitate sounds and follow a rhythm with their bodies, but humans tell stories and reveal their emotions by dancing. Maybe this is due to the size of our brain and its infinite potential to wire nerve cells in ever new connections, and thereby to learn.

Unlike any other species, we can move our bodies with precision and store in our memory an entire learned sequence of movements. Our fine motor skills are unique. Think of a jeweler who can create the most delicate filigree work. Our nearest relatives, chimpanzees, can grasp small things with their pincer grip, but have you ever seen a chimpanzee make jewelry? The precision of the jeweler's craft—being able to manipulate tools millimeter by millimeter—is beyond them. What's more, our fine motor skills—unique in the animal kingdom—are not limited to the movement of our fingers. Our facial muscles, too, can produce the briefest, most delicate expressions, so-called microexpressions, and our bodies can perform highly complex motor skills, such as when ballerinas dance *en pointe*. The miracle of a human being!

Isn't it fascinating what dance does to our bodies? The movement trains our muscles, while the music affects the messenger substances (hormones, neurotransmitters, etc.) that put us into a good mood or make us recall memories. Without us noticing, dancing changes us, as MRI scans prove.

We've been talking for hours, and although the dance floor is emptying, we are still glued to our bar stools.

"Enough of the theory! Let's do something for our neuroplasticity!"

We dance into the early hours of the morning.

5

PRESCRIBE DANCE, NOT DRUGS

••••••••••••

Wherever the dancer steps,
a fountain of life will spring from the dust.
RUMI

••••••••••••

It is the fifth day of the conference. Today's topic is interoception and movement, and we speak about what it is that goes wrong when we don't feel well. We get the sense that the science we pursue by day and the dance we indulge in by night complement each other beautifully—even if we aren't getting quite enough sleep. Today, though, we feel fit and alert. Last night, one of our colleagues was worried that he might be coming down with a cold, but he's fine now. He summarizes his recovery beautifully: "I must have danced the bug out of my system."

THE EFFECTS OF DANCE ON OUR HEART AND IMMUNE SYSTEM

Mikhail Baryshnikov is a famous Latvian American ballet dancer. He was particularly active during the 1970s and '80s after fleeing the Soviet Union, where he had been prevented from pursuing his passion for dancing the way he wanted to. In early 1998, Baryshnikov performed "HeartBeat: mb" at New York City Center (you can find it on YouTube). The "music" for this solo dance was his own heartbeat. With the help of a complicated ECG setup, Baryshnikov's heartbeat was played through loudspeakers for the audience; he literally danced to his own music. At times the rhythm of his heart so fired him up that he became short of breath; at other times it carried him along gently. His dance made headlines.

A very personal story had inspired choreographer Chris Janney to create this piece. His father had died from a heart attack, and choreographing this dance was Janney's way of coming to terms with his loss. For Janney, the heart is the seat of the soul. As biological research now shows, this is not such a far-fetched idea. What our heart does and how healthy it is depends to a large degree on how we feel.

The Persian philosopher Rumi wrote, "Wherever the dancer steps, a fountain of life will spring from the dust." When we watch people dancing, we typically sense an enormous energy and joy; dance is movement that comes straight from the heart. The first sound we ever hear is our mother's heartbeat. Because a fetus can hear the rhythmic beat of its mother's heart throughout pregnancy, it is soothing for babies to fall asleep on their mother's breast to the familiar rhythm. Similarly, many calming pieces of music and lullabies are written for a tempo of roughly sixty to seventy beats per minute and thus imitate the adult heartbeat.

Anyone providing first aid in an emergency is well advised to think of music too. The song "Stayin' Alive" helps to find the right rhythm for chest compressions. According to medical research, a frequency of 100 to 120 compressions per minute achieves the best blood flow after a cardiac arrest in an unconscious patient, and "Stayin' Alive" has 100 beats per minute. In his entertaining lectures, science slammer and cardiologist-in-the-making Johannes Hinrich von Borstel recommends performing cardiac massage to "Highway to Hell" instead: its 116 beats per minute make it even more suitable, although we may want to forgo singing it aloud, given the situation...

Our heart sets our rhythm; it is our pacemaker. It reacts with lightning speed to any change in our body. Whether we are happy because we succeeded at something or sad because something terrible has happened, our heartbeat changes. Sometimes a thought is enough to make our heart race, as anyone who has ever fallen in love can attest. Our heart puts us in tune with any given situation.

Different heart rhythms set off, or slow down, different metabolic processes. The impulse comes from the sinus node—a small muscle that sits on the heart. The sinus node is a veritable conductor. It can boost or slow down the immune system, and release hormones and neurotransmitters.

Our heart is our motor, continuously working for us whether we sleep, work at our computer, run a marathon, or dance. It beats about three billion times in the course of a lifetime and pumps 250 million liters of blood through the body. Probably no other pump in the world is that efficient. Taking good care of this motor is extremely important. Ischaemic heart disease and stroke are among the most common causes of death in the world.

Our heart is a muscle that we can train, just like any other muscle. Regular endurance training makes it stronger and larger and improves its blood flow, making it more powerful. A well-trained heart pumps more blood through our body with fewer contractions and is hence more efficient.

When you think of endurance training, what comes to mind first might be jogging, cycling, or swimming, but you can safely add dancing to the list. Its fast, rhythmic movements speed up our pulse, breathing, and heartbeat and, as a result, improve our endurance and stamina. Dancing has the added benefit of using almost every muscle in the body. Dancing truly makes our heart sing!

Music further amplifies this effect: Luciano Bernardi of the University of Pavia found in several studies that the tempo as well as the volume of music directly impact our cardiovascular system. Music that becomes continuously louder, in a so-called crescendo, stimulates our heartbeat. If a piece becomes quieter, in a decrescendo, our heartbeat slows.

In the early 2000s, Dafna Merom and her colleagues at Western Sydney University in Australia interviewed forty-eight thousand people aged forty and older about their hobby activities. Twelve years later, in 2016, the researchers reported their findings in an international journal. The result was unequivocal: people who danced in their leisure time had a significantly reduced risk of dying from a heart attack or stroke.

If we dance regularly, our health benefits more than from regular participation in any other sport. Why might that be? According to the Australian research team, it's because dance not only revs up our muscles and cardiovascular system but also has important effects on our brain; these, in turn, have a positive effect on our hormonal balance.

There's one particular swing dancer I really enjoy dancing with. One day I asked her about her secret. "You dance as if dancing is your life!" I said. She laughed. "For me, dancing really is the elixir of life. I am a cardiologist, and there's simply no sport that's better for your heart than dancing. That's why I tell my patients they should learn to dance. In my experience, having people go for regular weekly dances is one of the most effective ways of strengthening their heart."

Dancing may also offer promising rehabilitation opportunities for people who have already developed heart disease. In a study by medical doctor Romualdo Belardinelli and his colleagues at Ancona Hospital in Italy, 130 patients who had been admitted to the hospital with heart attacks were divided into three groups. One group went through the conventional cardiac rehabilitation program with treadmill and cycling; the second group attended dance classes that involved dancing the waltz three times a week in sessions of twenty-one minutes each; and the third group did not exercise at all. The results showed that groups one and two both improved in their fitness measures. Heart and breathing rates improved, as did oxygen absorption. The dancers, however, showed additional far-reaching improvements: they had improved their fitness measures and their arteries appeared more flexible. These patients were also in a better mood overall and reported improvements in their sleeping patterns, even their sex lives!

Endurance training is incredibly important when recovering from a heart-related illness, but motivation tends to be a big problem for these patients (and others too!). Seventy percent of heart patients drop out of conventional rehabilitation programs. The number is markedly lower for patients who dance—dancing is simply more fun. And there are benefits to dance when compared to other cardio exercises: we don't have to squeeze ourselves into proper fitness clothes, brave the elements, or shiver in narrow changerooms at the local pool. You can dance in your living room! Just play your favorite music…

When we dance, we do plenty of good things for our body at once: we train our heart and muscles, and we regulate our immune system. People who dance regularly fall ill less often. Our body gets attacked by germs during flu season, of course, but we're often exposed to viruses, bacteria, and fungi in our day-to-day life. It makes sense, then, to strengthen our immune system by eating a healthy diet and getting enough sleep. Getting enough exercise is, of course, also crucial. Many scientific studies have shown that moderate exercise boosts our circulation; by pumping more blood through our vascular system, the heart activates our immune cells. The number of so-called B lymphocytes increases. These blood cells are responsible for the production of antibodies and therefore play an important part in our immune defense. The more B cells we have, the more antibodies we can produce and the better we can ward off disease. In addition, the number of natural killer cells increases when we exercise. Killer cells are cells in our immune system that are able to recognize and destroy changed cells. Changed cells can be infected cells or even cancer cells.

Another positive aspect of dancing is stress reduction. Countless studies have explored the effects of stress on our immune system and have reached the same conclusion: permanent stress weakens the immune system during both acute infections and chronic diseases. During stressful periods we have less in our arsenal to fight off viruses, bacteria, or other germs. That's why we are more likely to fall ill in such times.

One indicator of stress is an elevated level of inflammatory markers in our blood. In 2003, Dana King and her colleagues at the Medical University of South Carolina were able to demonstrate that dance is among the leading means of improving the body's stress response. King's study involved 4,072 people who were surveyed about their hobbies. Those who listed dancing and jogging as recreational activities showed lower levels of inflammatory markers than other participants. Hobbies such as gardening, swimming, cycling, or weightlifting did not appear to have the same effect.

BACK AND JOINTS

Lunch break in Greece. A line has formed at the buffet.

"The first time I saw you, I was sure you'd been a ballet dancer, Julia. You have such good posture."

"Yes, that stayed with me—probably because I started ballet when I was just five. The story is that I was always sitting frog-legged on the floor and my parents worried I'd end up with knock knees and bad posture. So they figured I should start dancing, and that's how I eventually ended up doing ballet."

The spine is an important part of our body. It keeps us upright, supports us, and carries our weight. The spinal cord runs through the vertebral canal and is a core part of our nervous system. The human spine consists of twenty-four vertebrae. Sitting between the vertebrae, the flexible, more pliable spinal disks serve as shock absorbers. It is thanks to them that our spine stays flexible and we can bend and stretch in all directions. Our spine is fairly long, curved in two areas, and, seen from the side, shaped like an S. The purpose of these curves is to cushion the shocks when we run and jump.

Nerves grow out of the spaces between the vertebrae and extend into ever smaller nerve fibers throughout our body, like the branches and twigs of a tree. When we move our back, we stimulate those nerves and hence the entire peripheral nervous system that reaches into every organ, muscle, and connective tissue in our body. In this way, the spinal cord connects the brain with the farthest corners of the body.

If we don't move regularly, our back stiffens, and slowly but surely the unimpeded communication of our nervous system becomes rigid.

We only realize the importance of our spine when it no longer functions the way it should. Anyone who has ever had back pain knows this—and knows from experience how even simple movements can become sheer torture. Studies suggest that up to 80 percent of us will experience back pain at some point in our lives. Often, this pain is caused by the stress and tension we hold in the body. Sometimes we can even see this tension: when we are stressed and under a lot of pressure, we walk in a stooped posture, as if we are carrying a heavy burden. Our shoulders are pulled up and forward, and our spine is hunched. Sitting in front of a computer day in and day out is not good for our back either. People whose work requires them to sit all day put too much strain on their back. Provided a doctor gives us the green light, the best treatment for a painful back is movement, even though this may seem counterintuitive. When our back hurts, each movement hurts. Who feels like forcing themselves to exercise with a sore back? It's a lot more tempting to camp out on the couch with a hot-water bottle and some painkillers—but that's not helping our back one bit.

Another problem with back pain is the compensatory postures we may adopt to avoid it (for example, slumping the body to one side to avoid pain on the other, or shifting the trunk forward and curving the neck or shoulders to take the pressure off a tender area). It affects the mechanics of our movements: our pelvis and back no longer swing freely. This, in turn, creates additional pain. Have you ever tried to dance away your pain?

In dancing, a natural, upright posture matters, with the emphasis on "natural." But sitting at our desks has caused our good posture to fall by the wayside, and now we can't pick it up without getting lumbago…

Metaphors can be helpful when we want to develop better posture for dancing and in general. As already discussed, movement and language are managed by similar brain systems, and we can affect one with the other. Instead of simply ordering, "Shoulders down, chest out," the teacher in a good dance class will use imagery. Australian ballet teacher and former dancer Janet Karin, Medal of the Order of Australia, for example, asks students to use their imagination like a tool to improve their movements. For instance, she might suggest they imagine inhaling the fragrance of their favorite perfume or tell them to "stand proud." If you follow such instructions, you draw directly on your brain's motor system to modify your posture without explicitly having to think about the movement. We have all internalized what it feels like to be proud. Or how we feel when we breathe in a beautiful fragrance. Our facial expressions become gentle, and our wrinkles disappear. Our shoulders relax and drop, and we breathe more freely, which in turn relaxes our sternum and diaphragm, the muscle we use for breathing. So, sink deep into the floor, instead of "bend your knees." Rise into the sky as if pulled up by your hair, instead of "walk on tiptoes." Fly like a bird, instead of "stretch your arms out to the sides."

Dance often works with such metaphors, and the images prompt our brain to send the correct commands to our muscles. As a result, we move much more naturally and with less strain on the body than if we tried to consciously control the various muscle groups to do the movements.

Jan Gildea and his team from the University of Queensland in Brisbane, Australia, did an experiment to determine whether such mental imagery could have healing powers for back patients. Gildea strapped the participants—ballet dancers with back problems—into a kind of harness that allowed him to measure the mechanical vibrations caused by the dancers' movements. Then he gave one group of dancers precise instructions on how to tense and relax their muscles. The other group was told to "grow like a tree" or "make a flowing movement." The mechanics of the movements by the latter group improved dramatically, and their pain diminished.

Research has shown that belly dancing is particularly good for the back. The specific movements of the torso activate and then relax precisely those muscles that trigger discomfort and pain in the lumbar region. In 2017, Tabitha Castrillon and her team at the University of Central Florida conducted a belly-dance rehabilitation program for women with chronic lower back pain. After only six weeks of training, the participants were fitter, less sensitive to pain, and hence less fearful of triggering new pain. It is such fears that push many back pain patients into compensatory postures that further immobilize them.

Regardless of the music you dance to—"Gangnam Style" while house cleaning or tango in dance class—adopting a naturally straight position will, over time, become much easier in your day-to-day life. When you dance, you train the multitude of muscles, ligaments, and tendons that, together, stabilize your spine. Your upper body loosens up; the alternation of compression and decompression keeps your discs elastic, and elastic discs make for a flexible spine.

As long as I can remember, there was always something wrong with my grandmother's knees. My parents drove long distances to buy special ointments and tinctures and talked to many specialists who wore large glasses and hid candy in their desk drawers for me. While I was chewing on gummy bears, I admired the many colorful and complicated pictures on their office walls that showed the knee and other joints. I didn't understand much of what the doctors said, other than that joints were obviously a complicated business and had to be kept well-greased, like door hinges. I always dreamed of giving Grandma a tube of magical lubricant so she wouldn't suffer when she climbed stairs. Today I know that no such miracle lube exists. But if Grandma were still with us, I'd send her dancing.

Dancing moves our spine—and many other joints: in the legs and feet, the neck, the entire pectoral girdle, and, of course, the hips—in new ways. So it's not at all surprising that researchers at Saint Louis University in Missouri found dancing to be excellent training for people with arthrosis. They examined thirty-seven residents of a seniors' home, mostly women with an average age of eighty years, who suffered from arthrosis and who complained about stiff and painful joints. The researchers divided the group in two. The first half danced the so-called Healthy Steps—a slow, rhythmic sequence of movements—for forty-five minutes once or twice a week. The other half served as the control group, and didn't have any dance course. After just three months the dancing seniors required 39 percent less medication for their joint pain than before. And not only that: they were able to walk faster again. They also had a lot of fun on their dance days, and their mood improved.

The general rule applies here too: even if we are hurting, we must not stop moving. In the old days, doctors recommended immobilizing a painful joint as much as possible. They used casts and splints and even advised patients to quit exercising altogether. Today we know better. Scientific studies have proven that the cartilage in a joint is "nourished" by the synovial fluid. Movement distributes this fluid evenly in the joint and "massages" the nutrients into the cartilage. Less friction means less pain.

One of the really nice things about dancing is that everyone can determine for themselves how large and expansive to make their movements and how much energy to put out. In 2011, German medical researcher Marcus Schmitt-Sody at Medical Park Chiemsee was able to demonstrate that dancing offers good rehabilitation for patients after hip surgery because it doesn't jolt the new joint as much as, for example, jogging.

Even if pain is present, it seems that dancing can be a true miracle cure for our entire musculoskeletal locomotor system.

WEIGHT LOSS

We are sitting on the terrace of our hotel, having lunch. Truthfully, it seems as if we're eating all the time! The food here is so incredibly good: eggplant salad, grilled meat and fresh fish, many vegetables, rice, and beans. And the dessert buffet doesn't leave anything to be desired either: ice cream, cake, yogurt with Greek honey... and in the seminar room small bowls with cookies, chocolate bars, nuts, and fruit. Luckily, we dance all night. At least we're shedding some of the extra calories!

In 2009, season nine of *Dancing With the Stars* aired in North America. Reality TV star Kelly Osbourne was a contestant that year, and she and her partner, Louis van Amstel, finished third. Kelly still considered herself a winner, though, thanks to the forty-two pounds she dropped while competing. In an interview after her elimination, she credited the show and her partner with changing her life. Over the course of that season, millions of viewers could see what research has long confirmed: dancing is a wonderful way to lose extra pounds.

In December 1989, the *Journal of Sports Medicine* published a study on burning fat by dancing. Henry Williford and his colleagues were able to show that low-intensity dance steps burn four to five calories per minute, while dance styles that involve large muscle groups burn up to ten or eleven calories per minute.

You may have heard about aerobic and anaerobic types of exercise. The terms refer to our body's use of oxygen during physical exertion. Our muscles require energy to work and move. The energy is supplied either without oxygen (anaerobic) or with oxygen (aerobic). The anaerobic metabolism will only burn carbohydrates and not fat, because we need oxygen to burn fat. The aerobic metabolism, however, burns carbohydrates *and* fat to provide us with the fuel for movement. If you want to kiss your love handles goodbye, you need to choose the right kind of sport at the right level of intensity. The rule of thumb says that in aerobic exercise your heart rate should be less than 140 beats per minute. Of course, the number varies from person to person because weight and fitness play a part too. (If you want to know exactly at what heart rate your own body switches from aerobic to anaerobic energy supply, check with a sports physician for advice.)

Dance is an aerobic sport—that is, a sport that revs up our ability to burn fat. The high speed of dances such as tap or hip-hop causes the production of fat-burning enzymes. These enzymes are like vampires that crave fat instead of blood. So let us celebrate the enzyme dance! In a one-hour dance class we burn on average 330 calories—that's as many as in thirty minutes of cycling. With Zumba, it's even more: up to 400 calories an hour. Competitive dancers have fitness levels similar to those of competitive runners and burn up to 800 calories per hour.

In 1980, Scandinavian researchers Ewa Wilgaeus and Åsa Kilbom were able to show that the fitness benefits of the fast Swedish folk dance hambo were similar to those of running on a treadmill. A few years later, studies by Lucia Cugusi and her colleagues into the effects of the Sardinian folk dance ballu sardu produced similar results: the dancers' heart rate increased to 80 percent of their maximum heart rate, and the dancers burned on average 10.5 calories per minute, which means that ballu sardu can be classified as extreme exercise. The traditional folk dance with its deep cultural roots thus matches current advice on a healthy lifestyle.

Another good thing about dancing is that we often don't realize how quickly time passes. We may have been moving for a full hour, if not the entire night of a party, and haven't even noticed. This, too, has big benefits, because in aerobic training we initially burn mostly carbohydrates. It takes about twenty minutes for the body to start burning fat. After about two hours of physical exertion, our carbohydrate reserves are almost used up and we begin to significantly reduce our fat reserves. Have you ever jogged for two hours straight? Not many of us can answer that question with a "yes." But it's a good bet you've spent more time than that on the dance floor at some wild party or in a club... And if you ever participate in a dance event spread out over several days—a tango marathon maybe, or a swing festival or salsa congress—you can practically watch the pounds melt away.

Unfortunately, many people have a notion stuck in their heads that one has to be particularly slender and graceful to be able to dance. It's a prejudice we should ditch as fast as possible because no other sport is as suitable for any body shape. Beautiful proof of this fact can be found online: type "ballet" and "Lizzy" into your search engine and you'll be rewarded with numerous links to sites about American teenager Lizzy Howell. They almost all include a video that shows Lizzy pirouetting fouettés as she dances the Black Swan. For a few seconds, she focuses all her energy. The calm before the storm. Then she takes off: Lizzy swings her raised leg with a whipping movement high into the air and transcends into weightlessness. The music carries her as she pirouettes again and again. She spins as evenly as clockwork and ends the series of turns with a double pirouette before landing confidently in her closing pose. She's nailed it! Effortlessly!

Fouetté pirouettes are among the most demanding movements in the ballet repertoire. They require technical skill and a considerable amount of courage and confidence (they've been the nemesis of quite a few professionals over the years). Some ballerinas have a real phobia about the movement. Not Lizzy. She simply does it. And she does it despite not being one of those sylph-like ballet girls. Lizzy is overweight. Yet the ballet culture prevalent on the world's stages has imprinted upon us the "ideal" of petite dancers. The pressure to conform to this "delicate ballerina" norm is enormous, and is reinforced by the audience's expectations. According to surveys, more than

16 percent of ballet dancers suffer from eating disorders, not least because of the erroneous but dominant belief that certain ballet movements are impossible with a body mass index of more than fifteen. Lizzy proves that the opposite is true. "It shouldn't matter how much I weigh, the only thing that should matter is my passion for dance," she said in an interview. She is right!

We don't have to start with fouetté pirouettes; dance classes for ballroom, salsa, or Zumba are terrific opportunities to discover a passion for dance. Or take the Brazilian martial-art dance capoeira for great endurance training. In 2014, Rossana Nogueira and her team at the Menzies Health Institute Queensland in Australia studied an extraordinary type of school sport: the CAPO-Kids program, a ten-minute daily program of capoeira dancing for children. After nine months of thrice-weekly training, the 151 children that took part in the study showed lower blood pressure, reduced weight, and increased muscle strength. It's a very promising fitness program that also works for overweight children, which is especially important in light of the growing problem of too many children moving too little and sitting too much in Western societies of today. In New York City, almost half of all schoolchildren are now overweight. Such alarming numbers caused Jeanette Hogg and her colleagues at Weill Cornell Medical College in New York to embark on an exciting project in 2014: sixty-four children from nine to eleven years of age participated in after-school dance training for one hour three times a week. They were taught sweat-inducing styles: swing dance, cha-cha, and hip-hop. The kids loved it! And not only that—after four months of the program, these kids had not only lost weight, but they had also improved blood pressure and blood sugar levels. Most of the children had changed their diet too. Half a year later, 65 percent of the children had maintained these positive changes and habits. Moreover, a third of the participants reported spending less time watching television and playing video games because they were now practicing the dance moves at home as well.

Body perception through dance is a valuable and often new experience, especially for those who are overweight. Within the realm of dance, you can achieve visible success, regardless of whether you carry a few extra pounds.

Considering the ways in which dance benefits our cardiovascular system, revs up our immune system, exercises our spine and joints, and even burns calories, we really ought to be able to get dance via prescription.

In any case, we agree to set a good example tomorrow and introduce a dance break for the time in between lectures. It's time to bring some salsa to our neuroscience, and to let our synapses dance too!

6

DANCING AS THERAPY

If life brings you to your knees, do the limbo dance.
ANONYMOUS

Today's program includes Julia's presentation. The experiment she presents focused on how music and dance movements are perceived by people on the autism spectrum. The presentation triggers a long discussion about the practical value of social neuroscience. Researchers in this relatively young field try to understand what happens in our brain during social interactions.

"We especially want to see what happens inside the brains of people whose perception functions differently from that of the majority. In that way, we may discover opportunities to help them overcome difficulties," Julia explains during our lunch break.

DANCING OUR EMOTIONS

While I was completing my master's degree in Palma, I occasionally worked as an assistant therapist for a small boy on the autism spectrum. Pedro was four years old, unable to speak, and seemed very far removed from everyone. We usually spent two hours together, during which time I practiced with him the everyday skills that neurotypical children develop naturally as they grow up. Pedro would eat only two kinds of food, and it took us therapists an entire year to teach him to eat a piece of apple. His motor skills were also not typical for his age. Therefore, we practiced throwing and catching balls and balloons, and I worked hard to teach him how to hold a crayon in his hand. But, at times, it was difficult even to make eye contact with him. Progress was very, very slow. We followed a carefully crafted therapy plan, but at the end of each session we were allowed fifteen minutes to practice whatever we felt was right. My choice was always to put on some music. I would take Pedro's hands and just start dancing. Then everything would change very quickly. Pedro would smile and spontaneously look me in the eye. He would skip around me and tiptoe forward and backward; in these moments, his motor-skill limitations were barely noticeable. For fifteen minutes, he was almost like any other little boy.

I know, of course, that my experiences with Pedro don't constitute a representative study, not least because we have no data to show whether those fifteen minutes that I danced with him on Tuesdays and Fridays when he was four years old led to a measurable improvement in any area of his life, then or later on. All

I have is my experience and my positive memories of Pedro and his joy during our dance sessions. They are enough for me personally to believe that those thirty minutes a week had a positive impact. I am too much of a neuroscientist not to recognize that the joy and the movement must have left a trace in Pedro's brain and led to the creation of new networks that may have improved his motor skills and social behavior, if only in a small way.

For a long time, it has been common knowledge that dancing is therapeutic. In the 1940s, the Dance Movement Therapy (DMT) program was developed in the United States. It has since become a recognized method used in psychotherapy and body therapy. We now have numerous studies that illustrate the effectiveness of DMT for a variety of illnesses—even if we don't know precisely how and why dancing has these effects.

One of the reasons why dance is so beneficial is that it has a strong impact on our emotional well-being. Elizabeth Torres, a professor at Rutgers University, has researched autism spectrum disorders for years. Her findings indicate that autism could be caused by a movement disorder of the brain—an entirely new theory. One of the most important symptoms of autism is indeed limited motor skills. As we have seen in the previous chapters, everything we do in life is movement. We express our intentions and emotions through movement, from facial expressions to whole-body movements. Only through movement do we come to exist, and as we grow up, we become who we are through movement. All of these movements are made by our brain's motor systems. Torres and her colleagues wanted to know what happens when these motor systems do not function properly, as is the case with autism. People on the autism spectrum often have difficulties understanding the feelings of others and correctly "reading" the people with whom they interact. Given how important it is to understand the expression of feelings through movement during a social interaction, Torres and her colleagues' approach to studying autism is incredibly exciting. And dance presents itself as the perfect means for using movement to learn about emotions.

Today, DMT is used in therapies for chronic pain, depression, stress, post-traumatic stress disorder, eating disorders, and even cancer. In 2011, Antonia Kaltsatou and her colleagues at the Aristotle University of Thessaloniki in Greece designed a study to test whether Greek folk-dancing could improve physical functions, overall strength, and mental state of breast cancer survivors. Compared to non-dancing participants in the study, the participants in the dance group showed better physical functions, noticeable muscle gain, and—very importantly—both a significant reduction of depressive episodes and a considerable increase in overall satisfaction with their lives.

However, we need to emphasize that the science of the effects of dancing is still in its infancy. Researchers agree, though, that in most cases dance therapy beautifully complements and supports conventional treatments, not least with life-threatening diseases such as cancer. A cancer diagnosis upends a person's whole life. Many lose confidence in their body and have difficulty coping with the changes brought on by surgeries, chemotherapy, or radiation. Dance offers an opportunity for coming to terms with one's body and for learning to accept it with its disease. Dancing together with people who share the same struggle often creates new courage and hope. Dance can help patients with a life-threatening diagnosis to temporarily forget their disease.

During my time as a researcher in Korea I got to know many swing dancers. One of them was Jean, whom I befriended and danced with a few times. I learned recently that she had died. What was truly shocking was that she had suffered for years from an advanced cancer but had not shared her diagnosis with the dance scene, so hardly any of us knew about it. For Jean, dance was an escape into normalcy–or rather, it allowed her to escape from *her* normalcy, which was tied to disease, pain, and doctor's visits. I admired her attitude: she danced to look death in the eye, and with each step she seemed to tell it, "Even if you will eventually take me with you, I am still doing what I love most. I won't give it up!"

One of the goals of modern palliative medicine is to give patients choices, to allow them to die with dignity, and to support them in realizing as many of their remaining hopes and dreams as possible. A 2015 study by Sarah Woolf and Pam Fisher shows that dancing can be a good antidote to feeling helpless and sick. During a time when many patients feel dependent on their medical devices and medications, dancing restores some of their dignity. Incidentally, dancing is often more effective as therapy for the body than physiotherapy.

A whole range of different approaches to dance therapy exists, and specially trained dance therapists work in a variety of settings. Marion Spors at the University of Göttingen was able to show in 1997 that dance therapy was useful in helping patients to figure out the causes of their eating disorders. Spors describes how dance therapy allowed them to remember and identify abuse that had occurred within their family environment; victims of abuse found it easier to express their feelings through dance than to talk about their experience. In addition, trauma patients tend to suffer from some degree of alienation. If a woman or girl has been subjected to abuse, she tends to experience her body as an alien entity that is separate from her inner life, her feelings, and her core identity. Dancing can allow such a person to experience their body and soul as whole again and thus let them find their way back to themselves.

Even if we don't have to work through terrible traumas, dance can help us to maintain or regain our inner balance. "None of the world's books will bring you happiness, but they secretly send you back into yourself," said writer Hermann Hesse about literature; his words apply equally to dance.

Dancing is not only a good opportunity to express emotions, but it is also, as discussed, a wonderful method for "trying on" emotions. The dance floor is perfect for that: on the floor we can feel whatever we like, with no consequences for our "real life." The "letter exercise" in chapter 1 showed us how we can visualize many different emotions. We can switch from feeling sad or furious to feeling happy or cheerful in a dance, and adolescents might feel affirmed while posing and looking important during a hip-hop song. When we dance, we can practice being our authentic selves. Taking all its aspects together, dance can improve our mood and confidence, and along with this, our physical and mental health.

> A few years ago, a friend and I were booked to present an interlude in a show: a tango. At the time I had no experience with tango dancing, and I didn't want to take it on. "I'm a blond Scandinavian ballet dancer with a straight back. What do I know about Argentine tango?" I found the pretentious posturing of tango dancers silly and off-putting—too emotional and somewhat corny. My dance partner disagreed. "You're a dancer. You should be able to dance anything." That

was a challenge I couldn't resist. Then the rehearsals started. While I tiptoed across the stage nimble-footed and with perfectly controlled movements in ballet style, the choreographer kept shouting: "More passion! More feeling!" The list of critical comments was long but didn't help. Nothing did. One day I went to see my old ballet teacher and told him how discouraged I was by the whole process. He looked at me thoughtfully and asked a single question: "Have you ever been left by someone?" I thought about it, and all of a sudden, I felt again the loss, pain, rage, and sadness from that time.

The next day the choreographer had us dance the tango from start to finish for the first time. "Why didn't you dance like that right from the start?" he said afterward. I felt relieved and incredibly liberated, not because of the choreographer's verdict but because of the dance itself. In that moment, I realized I was capable of translating sadness into movement, which, paradoxically, gave me a feeling of liberation.

In an impressive series of studies, the Israeli neuroscientist Tal Shafir explored the influence of movement on our emotional state. As we have seen, movement originates in the areas of the brain that are closely linked to language and meaning. A body's physical movements can thus be "translated" into feelings.

Do the following exercise without thinking too much. Stand straight and let your body move as follows:

(a) Hand into pocket—grab key—low energy

(b) Chest out—arms up in a V-formation above head—inhale slowly—skip repeatedly in place

(c) Fix eyes on spot on floor—slowly duck down a bit—abruptly move backward—away from the spot

(d) Move forward—throw a punch—strong, direct, and sudden

(e) Core of the body droops—shoulders slouch forward—weight, heavy—passive

When we adopt these positions, we feel as if we really experience the situations described. When we put our hand in our pocket to reach for our key, the everyday act triggers barely any emotion, but in the pose of a victor with a puffed-out chest we feel strong and elated. Similarly, we automatically feel fear (c), rage (d), or sadness (e) when we adopt the respective poses. You didn't feel anything? Now that you know what emotions to feel, do the exercise again. This way you can practice how to trigger certain emotions in yourself.

Next, remain seated and simply imagine doing the same exercise. Visualize how your body will move. If you do the exercise repeatedly, merely imagining a movement will automatically stimulate your brain in a kind of feedback loop and activate the matching emotion. In 2016, Tal Shafir and her colleagues showed combinations of words like this to eighty research participants and asked them to do one of the following: adopt the poses, visualize adopting the poses, or watch short video clips in which other people adopted the poses. Each had the same result: the participants experienced the poses they adopted, imagined adopting, or watched others adopt as an emotion. Fascinating, isn't it? With a movement that you only see in your mind's eye, you can unleash an emotion as powerful as joy or rage.

This particular study and others like it are promising leads when it comes to developing therapies that use movement to tease out emotions or feelings you can "try on," in a manner of speaking. Dance movements are particularly suitable for such approaches because they can tell complete stories. In this way, dancing can heal. All you have to do is get off the couch.

DANCING AWAY YOUR INNER COUCH POTATO

Over the lunch break, it seems as if our team has lost some of its oomph. We've been talking nonstop, the warm Greek sun is making us dozy, and we're still feeling the aftereffects of last night's dancing. We'd love to skip the next presentations and go to the beach instead.

Dong fishes the afternoon's program from his pocket. "Two more lectures about social interaction and mental health. Surely, we can handle that! Let's go!"

His energy is inspiring. We all grab another cup of coffee and peel ourselves out of our comfortable chairs.

· · · · · · · · · · · · ·

During the photoshoot for this book, the makeup artist commented, "This dancing business is all well and good, but after my husband and I put our kids to bed we are bone-tired and simply collapse on the couch. Wild horses couldn't get us on the dance floor. Everything aches, and even going to the fridge feels like too much." I really can relate. Luckily, I have many tango friends in my life who, in spite of living in other cities or even on the other side of the world, are happy to remind me to go dancing. "It's only your head that's tired," they tell me, "not your body." When Friday night rolls around in London and I am still in my office at 11 PM, crunching data, when all my limbs ache, my eyes want to close, and my back is giving notice, I get shaken from my data-slumber by WhatsApp messages from Italy, Spain, Greece... : "Off you go!" For this is exactly the time when dancers head out to a milonga—a social event featuring Argentine tango that takes place all over the world every evening.

I peel myself out of the office chair I've been glued to for the past twelve hours and stretch. Every single part of my body screams in protest. "Go home and go to bed," says a voice in my head. But no. I am on a mission: to tango.

The need to gather up all of your gear is a sure-fire way to kill any anticipation when it comes to dancing. That's why I take care in the morning to pack a bag with my hairbrush, makeup, dance clothes, and shoes. Then I only need to pull out the bag to start the transformation from Research Cinderella to Tango Princess. At a milonga, there are only princesses and princes, nothing but splendor and glitter and an awesome red light that sometimes turns golden, as in fairy tales. Of course, my dance clothes are black and gold, and I even have golden eye shadow... Even my shoes are gold. As soon as I put them on, I feel transformed—and very different from when I'm in my lab runners.

Stepping out of my office into the hallway of the dark institute, I'm again beset by doubts—in spite of my glittering shoes and red lipstick. Even more so as I open the door to the street with my employee chip card, and the cold air of a London night hits my face. What on earth am I doing? Cold air makes our bodies contract toward their core so they won't lose too much warmth. We hunch our shoulders, which further reinforces the feeling of rigidity. No wonder the voice in my head is still telling me to go home! But I am a neuroscientist; I know my inner couch potato and how to outfox it. That's why I glance at the glittering tango dress that flashes below the hemline of my coat. The sight reminds me of many fantastic milongas of the past and gives me a touch of the warm feeling that only dance can produce. The breathless dance appears in my mind's eye and floods me with warmth against the cold of the night. It makes me think of the *Harry Potter* books: J.K. Rowling invented the Dementors— evil beings that suck the life out of us—but she also invented their antidote. With the help of a warm and good memory we can generate a shining Patronus, a protector with whose help we can make it through the darkness—and through the London rain on our way to a milonga. We typically store experiences that carry emotional energy more solidly in our memory than trivial everyday moments. Ask someone about their wedding day and watch what happens. Even if it occurred twenty years ago, they'll likely be able to tell you about the weather and the world events that were unfolding that day, and probably even share minute details of their wedding dinner. Sometimes a tiny spark is enough to light the fire again and motivate us. So I set off with my glitter-dress Patronus, and as soon as I turn onto the street where the milonga takes place, I hear the first sounds of tango music.

As the music becomes clearer, I quicken my steps. At the Negracha Tango Club, I barely have my coat off when someone invites me to dance.

I forget everything and let the music take us away.

From a biological perspective, enjoyment is a signal that uses messenger substances to communicate with our body. The brain releases the neurotransmitter dopamine to give us a happiness kick and thus make us repeat this behavior again in the future.

Our brain remembers the feeling of pleasure and connects it to the activity that prompted the pleasure. When we are hungry, we may suddenly imagine a plate of steaming spaghetti topped with our favorite sauce because that particular food gave us great pleasure before. We'll do anything to once again enjoy this dream plate of pasta. Experiences that gave us pleasure in the past become objects of desire and make us long to repeat the experience. The longing is triggered by the dopamine action in the brain's pleasure systems. Drug addiction, too, stems from the activation of those systems.

Dopamine is a miracle substance that motivates us and makes us act with intention. It is the brain's neural reward system that drives us. It was discovered in the 1950s by pure chance. American researchers James Olds and Peter Milner at the California Institute of Technology were studying the behavior of laboratory rats when exposed to adverse conditions, such as electric shocks. When the scientists inadvertently placed the electrodes in a different part of the brain on one of the rats, the animal kept returning to the spot where it had been given the electric shocks—even the following day. It was hoping for more shocks. To examine the phenomenon more carefully, the researchers placed rats in a box with a special lever with which the rats could shock themselves. After a few minutes of learning time, the rats stimulated their brains by pressing the lever regularly about every five seconds. They forgot to eat and pushed the lever until they collapsed from exhaustion. The researchers had discovered the brain's reward system and were able to identify dopamine as its main agent. No substance appears capable of motivating us more than dopamine.

The less dopamine is released, the sadder and more apathetic we become. But we can take deliberate action to get on top of that listlessness—namely, push the lever and increase the supply of dopamine in our brain. But don't worry; you don't need to give yourself electric shocks like the rats did. Instead—yes, you've guessed it—you can go dancing. To begin with, and as you will remember, dopamine is released when we listen to music. Dopamine levels also increase when we experience success or are well on the way toward it.

If we set our goals too high for ourselves, we won't get much of a sense of achievement. To get the dopamine kick that motivates us to continue, we are better off setting achievable goals—and the dance floor is a perfect place to do that. Every step counts. Are we focusing on a particular step sequence today? A turn? A particularly tricky passage? Every positive learning experience activates the brain's reward system. If we break down our long journey into small segments, we will have moments of success more frequently. And they, in turn, compel us to continue the journey. An elixir for our motivation!

Unlike, say, jogging, dancing doesn't require you to endure endless and lonely miles of pounding the pavement, sometimes in less than ideal weather conditions; anyone who goes dancing joins a group or dances with a partner, which adds motivation. After all, we'd rather stand up the couch potato inside us than our dance partner. And if we have a lot of fun in dance class, we'll take those memories with us into our day-to-day lives. Each time we see, hear, or smell anything connected to that experience, we return, emotionally, to that happy space. The melody of a particular song can take us back to a cherished moment in our life. The smell of freshly cut grass is enough to put us into a summery mood. And a steaming cup of hot chocolate can bring memories of cozy winter days spent around a fire.

The next time we find ourselves lying on the couch feeling grumpy and lethargic, the music we heard in our last dance class or the sight of a tango dress may be enough to motivate us again.

That glittering tango dress and those fancy dance shoes have another effect as well. The desire to adorn our bodies and make ourselves look beautiful is a deeply human urge. Archeological excavations have shown that even our earliest ancestors painted their bodies and used beads and bones for jewelry.

In fact, current studies have proven that dressing up isn't just nice for others to look at but also has an impact on our own mood, motivation, and confidence. When we change what we wear, we change our body language and the effect we have on others. Studies about the use of makeup have shown a similar effect. Some women find it easier to present themselves more confidently when they use makeup than when they don't. Makeup can have a boosting effect on our self-esteem and hence on our motivation. But before you squeeze your feet into pumps and paint your lips, remember that it can be enough to simply exchange track pants for jeans in order to feel better and ready for action. How about a dance? Many terrors and Dementors lie in wait on the way to the dance floor: tiredness, stress, that inner couch potato, or the rain in Vancouver (or New York, or Montreal). But armed with an understanding of our body and a few good memories, we can ward them off. The reward is priceless: we feel, float, laugh, enjoy, and get motivated for the next day.

STRESS: BITTEN BY THE SPIDER

Dong has to slip out of the afternoon events twice when his cellphone rings. His research team back in Germany can't find a document that is supposed to be stored on the institute's server. It's difficult for him to provide help all the way from Greece, not least because he's presenting his own research tomorrow and has set aside today to review his images and diagrams. His time management is getting wobbly.

"It doesn't look like I'll be dancing tonight," he says. The stress he is feeling is evident in his voice.

Most of us are familiar with stress. The children are whining, the fridge is empty, the cellphone is ringing for the umpteenth time, and we really should get back to our desk ASAP.

Every one of us has experienced situations when it's all simply too much. At some point, nothing is working anymore. We feel stressed—which is not a good feeling and, in the long term, will make us sick. Yet stress is first and foremost a natural reaction to challenges, and it actually makes a lot of sense in evolutionary terms. When one of our forebears was faced with a saber-toothed tiger or an opposing tribe's most powerful warrior, they had to make decisions in a flash. Their brain declared a state of emergency—and our brains have continued to do this ever since, any time we feel threatened or overwhelmed.

When this happens, the stress hormones adrenaline, insulin, cortisol, and noradrenaline are released into the blood, and our entire being goes into a state of high alert. As we saw when discussing cases of trauma, this type of stress response affects the brain as well. We get tunnel vision. Don't expect thoughtful, creative solutions to problems at this stage; when we're face-to-face with a deadly snake or saber-toothed tiger we definitely don't need complex brain functions.

The stress hormone par excellence is cortisol. It is important and useful when we have to quickly escape from a dangerous situation—say, a snake in the forest—and it helps us in highly intense work situations to achieve peak performance. Cortisol speeds up our heart rate, pumps fresh blood into our muscles, and switches the body from rest mode to work mode. Digestion and regenerative processes slow down to let us focus and decide on a fight-or-flight response. We run from the snake, or we work through the night to meet a work deadline. Once the challenge has been met, the extra effort we expended takes care of reducing the excess cortisol in our body.

However, today's saber-toothed tigers tend to be more subtle, and nobody runs away from an endlessly ringing cellphone, a jam-packed calendar, or a nagging boss with control issues. Our body produces stress hormones in response to these things, but then the tension remains. Our fight-or-flight reaction—which would typically neutralize our heightened alertness—never comes about, and the extra hormones released into the blood have nowhere to go.

Sometimes stress paralyzes us. That's because the brain region responsible for survival can make us rigid with fear. We hold our breath and freeze. Through the nerve fibers in our spinal cord, the brain has direct access to our leg muscles, which is how stress can rob us of all muscle tone. This was not a bad survival strategy when an enemy in days gone by was either too fast or already too close—a "playing dead" response. You may have experienced a less extreme version of this reaction at some point in your life, if you found yourself in a scary situation. Perhaps you felt your knees go soft, and you seemed to suddenly lose all your strength. Neither response is deliberate; they are the body's physiological reaction to an overwhelming threat.

If stress continues for a long time, it will make us sick and can lead to utter exhaustion and burnout. Typically, the first symptoms of burnout syndrome are fatigue, sleep problems, and head and stomach aches. Because the immune system is no longer working optimally, we are at an increased risk of illness. Being permanently on high alert also causes high blood pressure and, as a result, leads to a greater risk of stroke and heart attack. Clearly, it's very important to reduce stress!

Our primal instinct reveals the ideal reaction to stress: run as fast as we can. Or fight. But we prefer to relax on the couch. Many studies show, however, that movement helps to reduce stress and releases endorphins. We have had scientific evidence since the 1980s that aerobic activity clears the blood of the stress hormone cortisol.

This is another situation where dance can help. People have always known this—even without scientific studies. A good example is the tarantella, a folk dance from southern Italy that takes its name from the wolf spider. The dance seems hysterical because it imitates the fever and agony of a person who has been bitten by the spider. With legs twitching just above the floor, the dancers seem to glide through the air. In the old days, the dance would sometimes continue for days, with people dancing themselves into a trance to eliminate the "poison" from their bodies. In her book *Dance and Health: Conquering Stress*, the anthropologist Judith Lynne Hanna suggests that in the tarantella the spider's poison is emblematic of difficult times, interpersonal conflict, and similar stresses. Thanks to the convulsions and the exhausting duration of the dance, a person would sweat and thus rid themselves of any form of "poison." Their stress was diminished, and their mood improved. To this day, Neapolitans use the tarantella for the same reason: in response to natural disasters, terrorist attacks, or political injustice, they dance the "poison" out of their bodies. The Italian folk dance has not lost its significance as a healing ritual.

I can well remember the cold November night in London when friends took me to a tarantella evening. It was the tail end of a week of long hours of overtime work, and I was pretty stressed out. Because I was not familiar with the tarantella, I googled it. "Ecstasy through dance" and "you'll forget everything around you" sounded promising! At the club that night, the band didn't perform on the stage but played, danced, and sang right in the middle of the crowd. The music was reminiscent of summer and vacations in Italy, its rhythm electrifying. Without a second thought, I flung my bag and coat into a corner and threw myself into the fray. Amidst grooving and swinging bodies, I was gripped by the group's energy. My legs just followed the unfamiliar rhythm. The steps didn't really matter; what mattered was not to miss a beat. In no time my heart was racing, and I was taking deeper and deeper breaths. I forgot the stressful week and the chilly London fall outside; it felt as if I'd shed my armor. At the end of the evening, it seemed to me that I'd burned up more energy and was more out of breath than ever before—which is saying something, given that I'd done ballet for years and had had many hours of pretty intense training!

After this wonderful evening I decided to sign up for a whole tarantella course and to dance the tarantella whenever I get bitten by the "stress spider."

If the tarantella isn't your thing, what's stopping you from waltzing through your kitchen or doing hip-hop in the hallway? Embarrassment? Only the first time!

As part of the research for her doctorate at the Goethe University in Frankfurt, Colombian psychologist Cynthia Quiroga Murcia analyzed the saliva of twenty-two couples before and after dancing tango. She measured their hormone levels and asked the participants to fill out questionnaires about how they felt. The results matched what many people experience after an evening of dancing: dancing lowered the levels of the stress hormone cortisol and in both partners increased the level of the sex hormone testosterone. To determine whether these positive effects were caused by the music, the movement, or the physical contact between the partners, Murcia and her team tested the participants for each of the three factors separately. It turned out that the reduction of cortisol mainly results from the music, while the release of testosterone is caused by physical contact and shared movement with a partner. When all three factors came together—as when dancing tango—the effects were strongest.

In 2004, Jeremy West and his colleagues at Reed College in the United States compared the health benefits of African dance and hatha yoga. Three groups of students participated in either one hour of African dance or one hour of yoga or listened to a biology lecture. Afterward, both the dance group and the yoga group reported a significant improvement in mood. All participants felt less stressed. But it was exciting that the researchers were also able to observe a significant difference in the concentration of messenger substances in the participants' blood: cortisol levels rose in the African dance group and fell in the hatha yoga group. As both groups reported lower stress and better mood after their respective class, the researchers concluded that people can derive different benefits—depending on what they need—from the two types of exercise. After all, a higher cortisol level in one's blood is not only indicative of stress but also of the positive effects of an appreciably higher level of energy and joie de vivre.

Dancing can also have a meditative effect because step sequences and combinations of movements demand and foster concentration. This makes dancing a particularly good choice for people who have difficulty with silent meditation. Yoga, qigong, or tai chi have become well known as forms of meditation that are not practiced while sitting quietly but lead their practitioners to relaxation through focused, mindful movement. Not equally widely known or practiced as yoga or qigong, but similarly effective, are meditative dances. They foster our power of concentration, coordination, flexibility, and memory. In dance meditation, steps tend to be very simple and are often performed by a group dancing in a circle. The goal is to find inner calm, gather new strength, and be mindful and aware of one's body.

Since it has been well documented that mindfulness meditation reliably leads to relaxation and improved mood, researchers at Duke University in the United States wanted to find out whether African dance can have such effects as well. For eight weeks two groups of participants attended an hour-long class of either ngoma dancing—a dance style from Tanzania—or mindfulness meditation. The before-and-after measurements were unequivocal. After the eight weeks, both groups showed fewer symptoms of anxiety or depression and reported a higher quality of life and stronger social bonds.

The idea that we can use rhythm to help train ourselves to concentrate and to connect with our inner selves can also be found in TaKeTiNa. In this dance meditation, people move to simple rhythms while simultaneously repeating rhythmic mantras. Because research has demonstrated its effectiveness for both our physical and mental health, TaKeTiNa has become a very useful tool in psychotherapy and pain management.

Moving to a rhythm—whether wild and fast as in the tarantella or mindful and focused as in TaKeTiNa—can work wonders against stress: step by step, or with a two-step, it counteracts stress.

> Even if we sit all day, we can suffer from enormous, horrible stress in the absence of any visible threat. In the crazy competitiveness of the academic world, stress comes mainly from university policies that require faculty and researchers on short-term contracts to constantly publish new work in high-impact journals if they want to have even a tiny chance to keep their jobs. The hours of the day slip through our fingers while mountains of work pile up in front of us. How can we ever hope to get on top of it all? The pressure stresses me out. I am aware that in this race with the Usain Bolts of the science world, my body has probably been releasing cortisol into my blood all day

long and has hence been on high alert for hours on end. Also, my muscles are paralyzed by the pressure of this anxiety-inducing situation. There's only one solution: get out and dance!

AGAINST ANXIETY

"Take a deep breath, Dong. Getting yourself all worked up doesn't help anybody."

Right now, it's difficult for Dong not to become frantic. His colleagues have still not been able to locate the badly needed documents.

We talk about relaxation techniques. Dong swears by qigong, a Chinese combination of physical exercises and meditation. We go down to the beach, take deep breaths, and Dong demonstrates some of the simple exercises. In qigong, you move slowly to the rhythm of your own body, as in a kind of meditative dance.

"Of course, it only helps with ordinary stress," Julia adds. "When the going gets really tough, you need more, and different, support."

Julia knows what she is talking about.

．．．．．．．．．．．．

When we are in imminent danger, seconds suddenly stretch out and time appears to pass more slowly—as did the milliseconds between the moment my foot lost its bearing on the stairs and the moment I crashed onto the floor below. I could hear the sound of bones breaking in my body. It was terrible. I was entirely in the here and now and able to think clearly. Important episodes in my life moved past my mind's eye, one image after another—like in a flip-book when your thumb is too slow. In that instant I sensed that my life was going to be different. Then I passed out.

I had fractured my sacrum and my elbow. I was bedridden for a long time, and even moving the short distance to the bathroom was very hard. Recovery was a long journey, which gave me time to bid farewell to dancing and to the body I'd once had. I knew it was the end. Many dancers' careers end this way, or in some similar fashion. I tried to convince myself that my case was nothing special, but to me it felt like the end of the world.

Of course, it was not. The world kept turning. It simply kept going—and along with it, so did I. In time I came to understand that what I'd seen as the end, period, was really only the end of *one* world. It was the end of the world in which Julia would become a dancer. But what was much more important was that it was also the beginning of an entirely new and exciting world. A world in which I found a new mantra for myself: if I can't dance with my body, I shall dance with my mind.

Remembering the sensations connected to a traumatic experience arouses such strong feelings of discomfort that we simply cannot bear any of the smells, sounds, or other sensory phenomena related to the trauma. A traumatic experience overwhelms the brain's normal mechanisms for coping with stress. The memories are split, in a sense, and stored in different systems of the brain. The trouble with this is that no temporal allocation is coded for a trauma. Months or years later, smells or sounds can still act as a trigger and leave us stuck in a frightful memory, as if the trauma were happening all over again. It can feel as real as it did the first time.

Our brain does not care what it is that triggers a trauma; once a trauma has occurred, it is there and will continue to provoke similar reactions. Whether we survived a plane crash or only believed, for a brief moment, that our loved one lay crushed underneath a collapsed house, traumas are always overwhelming, terrifying, and distressing experiences that leave us feeling incredibly helpless. For a moment, we have lost control. In general, we cope quite well with such stresses, which is why so many traumatic experiences don't require therapeutic help. Problems arise only if the trauma affects a person's everyday life. The diagnosis in that case is called PTSD, or post-traumatic stress disorder.

Most trauma patients suffer from the anxiety and helplessness caused by the loss of control they experienced. As a result, they are constantly under pressure and tend to develop grave physical symptoms that can seriously impact their day-to-day living. Often the hardest thing for people experiencing anxiety and panic attacks is explaining to others why they are suddenly afraid and have withdrawn from their environment. It is important to understand that during a panic attack the same biochemical processes take place as during a genuinely dangerous situation. Cortisol plays an important role and puts the entire body on high alert. No wonder trauma patients often have an elevated heart rate, high blood pressure, and suffer from stomach problems, headaches, and chronic muscle tension.

These symptoms, together with the anxiety itself, cause many traumatized people to no longer experience their body in a positive way. A whole range of studies show that people with an anxiety disorder have increased interoceptive accuracy. Remember chapter 4? Patients with anxiety are more alert and are always on guard. Because they sense their body so intensely, they are on the lookout for any physical reaction, no matter how small, and they are ready with an interpretation of its meaning. If a racing heart can signal a panic attack, I will pay special attention to my heart and my pulse and will likely worry that any change is indicative of a threat: "I'm in danger." After experiencing a trauma, we have to relearn how to deal with such feelings and sensations.

Here, too, dance therapy can work small miracles. When we dance, we concentrate totally on ourselves. Emotions and movements form a whole and, as a result, help us to develop better body awareness.

"To stand on one's own two feet," "to get cold feet," "to pull the rug out from under one's feet"—many such idioms reflect that we must be able to trust our feet to feel safe and secure in our lives. Anyone who has ever had wobbly legs and been fearful they might not provide the required support knows how uncomfortable the experience can be. Dance therapists operate on the idea that we strengthen our self-assurance and confidence by experiencing that we are in control of our bodies and that we can walk with firm steps—through dancing. Ideally, the new neural connections that we establish by having this feeling of safety repeatedly will lead to improvements in other areas of our lives.

After my accident, I no longer felt at one with my body. I felt very insecure, and I definitely didn't have "my feet firmly planted in my life." As we saw earlier, our sense of identity heavily depends on our familiar movements and sensations. In a way, we are what we routinely do and feel. And I, I no

longer danced. Before my fall, I had trained every single day, but now, suddenly, all of that was gone: the music, the movements, and the emotions attached to them. Inside, I was moving away from the Julia I had been before. This was not just a psychological process but also a physical one in the neural networks of my brain.

But then, one night, years after I'd injured my back, my friends dragged me out to a disco. That evening turned out to be healing on many levels. We danced the pony farm dance and whatever else came to mind. Without thinking, we were pure emotion and entirely in the moment. The dance moves of that night suddenly reminded me of my former self; my neurons were back in full swing and firing in patterns they hadn't for years. It felt as if, at long last, all my neurons were finally communicating again with one another. The world kept turning, the music kept playing, and I could sense tears in my eyes. It was a feeling of liberation I will never forget. I was able to dance again.

Such an experience can start the process of recovery. Dance stimulates our awareness of our body; we suddenly feel our heart beat again—and with it the feelings that can guide us.

DEPRESSED? DANCE IT OUT!

The day's presentations are done, and we are now settled on the sandy beach, watching the waves. "Dancing makes people happy, but my own story of dancing begins at the opposite end of the emotional scale."

Dong is talking about his depression. About the dark days when he would have preferred to hide in bed. When he didn't want to see anyone.

"The longer you are in that state, the more lethargic you become. You feel stuck in a black hole. It sucks up all the world's light; everything keeps getting darker, and it's almost impossible to climb out of it under your own steam."

"How did you manage?"

"I had a good doctor—and a wonderful childhood friend who talked me into swing dancing, into doing the Lindy Hop."

We all know what it feels like to be sad. Sadness is part of life, part of being human, like happiness. For many of us, periods of sadness and melancholy are times of intense reflection about ourselves and our world and, as a result, can be times of great creative output. Hardly another emotion has contributed as much as sadness to the development of our society's art and culture.

When we lose a loved one through separation or death, or when we are disappointed in or rejected by someone we care about, it is natural and healthy to be sad. Sadness originates from a cocktail of messenger substances and hormones; for example, our blood's serotonin and prolactin levels drop. Serotonin is a neurotransmitter that maintains order in both our brain and our digestive tract. It regulates our mood, appetite, and sleep pattern. The release of this messenger substance is reduced when we grieve, which in turn makes us stressed and paralyzed. One clear sign of sadness is catatonia: remaining in an immobile posture and staring into nothingness. Another sign is the convulsed exhalation when we cry. We are suffocated and consumed by our own lack of action, and this void threatens to shatter us. Over the years, studies have shown that our emotions flatten when our body is immobile—people in this state describe that they feel like hollowed-out zombies.

If we are stuck in a depressed mood, we can change how we feel by increasing our serotonin. There's a reason most commercially available antidepressants contain serotonin. But can we boost our body's production of serotonin without medication?

In 2005, Korean neuroscientist Young-Ja Jeong published a study about adolescents suffering from mild depression. The researcher checked the participants' mental health and any changes in their neurohormone levels after twelve weeks of dance movement therapy. She was able to show that serotonin levels in the participants' blood had markedly increased. A meta-analysis by Sabine Koch at Heidelberg University confirmed that DMT generally improves participants' mood.

Tango dancing, too, appears to help with depressive moods. A 2012 study with people suffering from depression compared the effects of tango dancing and mindfulness meditation. The participants who took part in the tango dancing class described significant stress reduction and distinctly greater life satisfaction.

Have you ever watched an episode of *Grey's Anatomy*? A recurring theme in the show is that two of the main characters, Meredith Grey and Cristina Yang, like to dance out their inner conflicts. Whether dancing in the operating theater or in their underwear on a bed, their motto remains the same: "Shut up! Dance it out!"

Dance also appears to have a positive effect on our mood during puberty—a time in life when we often feel blue. Some girls become extremely shy during puberty. They tend to suffer from depressive moods, low self-esteem, and psychosomatic symptoms such as stomach pain. In one clinical study, Anna Duberg and her colleagues at Örebro University in Sweden examined whether a dance class might help. The teenagers loved the class and mostly judged it very positively. In a questionnaire after the class was over, the girls who had participated described their health as better than did a control group of girls who did not attend a dance class. So, if you have a grumpy teenager in your house, try sending them off to dance! There are also dance apps and online dance classes available from around the world, at any time of the day. Want to dance Argentine tango in your kitchen? Try out a Persian dance class to light up your endless Monday afternoon? Virtual dance classes abound—it's up to you to pick and choose to your heart's content.

A few years ago, a friend and I produced a show for a dance theater troupe. I was responsible for the music, and Jeannette for the dance. As we were working on the show, Jeannette lost a loved one—a deep shock for her. One afternoon, her pain had become unbearable. The stage was pitch-black and empty. It occurred to me to play one of my compositions on the grand piano. Without thinking about it and without speaking a word, Jeannette started to dance. She expressed in her dance how she felt inside. All I could make out on the stage was her shadow, but I sensed what was going on inside of her, and I tried to absorb her pain and grief and translate it into my music. As if in a trance, I wove those feelings—feelings that weren't mine—into my music, and we became one during that magical moment. A dance at the piano, a musical improvisation on the stage. While we were playing, I, too, started to cry as I pounded the piano keys.

I have no idea how much time passed, but at some point, Jeannette lay panting and sweating on the stage floor. We were both bleeding—my fingers from pounding the keys and Jeannette's feet from dancing barefoot—but neither of us had noticed.

What had happened? The extraordinary emotional burden—a combination of psychological stress and physical exertion—had, in each of us, triggered a release of endorphins that had numbed our pain. For a moment, we had danced away the pain.

It is generally very helpful when dealing with stress, despair, and emotional pain to do things that automatically trigger the release of endorphins and serotonin in our brain. It will at least temporarily dull the sharpness of our pain and provide us with renewed motivation. Dance can also help on this front through its positive social aspects; people who dance meet others and leave behind their loneliness. As a result, the vexatious merry-go-round in our mind that can drive us crazy when we are sad will stop, at least for a while. Here, too, the "trying on" of emotions in a dance can be helpful. Dr. Lioba Werth at the University of Würzburg has coined the term "body feedback" in this context, and a growing number of studies have shown that movement can have positive effects on the emotional stirrings of our brain. For example, it is difficult to keep feeling grief and stress if we join in on a dance with open arms and life-affirming, powerful movements—and perhaps even smile as we do so.

Do you remember Tal Shafir's research into the "trying on" of emotions that we discussed earlier? We're dealing with the same issue here.

But crying, too, can help. The reason we cry when we experience grief or pain is not yet fully understood. We all know the sense of relief that comes when the dam breaks and we can finally let go and cry. That feeling is caused mainly by prolactin, a hormone that makes us feel comforted and is also released when we are with people we feel safe with and feel their closeness. One study showed that tears shed by research participants when they listened to melancholic music contained prolactin, but tears shed when they cut onions did not. The composition of emotional tears hence differs from that of tears caused by the reflex to something that hurts our eyes.

Add to that another curious fact about our brain: in the brain, the system dealing with pain and the one dealing with happiness overlap. Pain can therefore paradoxically be transformed into pleasure—if we know how to do it! This has been shown in research by Morten Kringelbach as well as in studies by Siri Leknes and Irene Tracey at Oxford University. They, together with many others, theorize that art and dance can be used as a kind of catharsis for emotional cleansing.

Have you ever felt blue and listened to the same sad song over and over until it made you cry? To the point where you were completely drained and wiped out but felt better anyway? When we do that kind of thing, something in our brain does seem to switch from "pain" to "happiness." It's worth remembering that there is a time for everything in life—weeping and laughing, mourning and dancing. It's a mantra we should embrace for the darker days life holds in store for us.

> It was one of those evenings when you simply can't bring yourself to do another thing. The day had been long and arduous. Nevertheless, Inga, Philipp, Julia, Linda, and Lisa managed to talk me into going out for a glass of wine. We talked much, and we drank much. It ended up being a nice evening–not least because it had been quite some time since Lisa

had joined us. Over and over again she had canceled on short notice, even though that wasn't really like her.

At some point, the topic of dancing came up, and Inga and I danced a Slow Lindy for them, and then a blues followed by a Charleston. Soon enough, we had talked the others into dancing too. We showed them the steps and how, with some simple moves, you can give yourself over to the music. Everyone joined in enthusiastically. Suddenly, in the middle of a dance, Lisa burst into tears. "I've got to tell you something." We looked at her. "I've lost all my strength and don't have any more energy," she said as tears ran down her face. "I'm close to calling it quits." We looked at each other in shock. Lisa had always been someone who laughed a lot, the person to whom everyone gravitated and took an instant liking. She had an impish look and a genial, warm manner that won over people's hearts. The deep sadness inside her had remained hidden from us, even though we, as psychologists, should have been worried by the symptoms of the past months: withdrawal, excuses, unanswered WhatsApp messages. "But," Lisa continued, "right now it's wonderful to see you and to dance with you. I suddenly feel there may be a light at the end of the tunnel. That I can actually be around other people and that there are good things in life."

We were very worried and stayed all night with Lisa. At dawn, Inga took her to a psychologist and doctor. She was prescribed appropriate medication and given a referral to a good psychotherapist.

Today she is well again. And she kept one very special thing from that evening: dancing. Since that fateful night, she never misses an opportunity to go swing dancing.

We don't think that dance per se can heal depression, burnout, anxiety, or trauma. Depression, for example, is a grave mental illness that must be taken seriously and treated professionally. While sadness or a passing listlessness are part of life, a depression does not go away on its own or improve in response to encouragement. In a depression, certain biochemical processes in the body are persistently disrupted. Depending on the severity of the depression, medication, psychotherapy, or a combination of both is required. Only a doctor can make that decision. But dance medicine can beautifully complement conventional medical treatment. One thing is certain: dancing won't do any harm.

IF ONE OF OUR SENSES IS MISSING

The sun has set, and we've somehow missed dinner. It's been a very thought-provoking evening, and we're both quieter as we enjoy the sound of the waves and the view of that fiery disk on the horizon that is spreading gold on the sea. Lost in thought, Julia is dancing a few steps on the sand.
"Just imagine being unable to take in any of this beauty…!"

In 2018, the Paralympic athlete Danelle Umstead was a contestant on *Dancing With the Stars*. Umstead, who suffers from retinitis pigmentosa, was the first blind contestant in the show's history. Many television viewers had a hard time imagining that a person could master complex choreographies without being able to see.

Some dance schools now offer lessons for the blind. In New York, the National Dance Institute welcomes children of all abilities, including those with vision impairment. In the Austrian city of Graz, the Conny & Dado dance school offers Blind Date Dances, events whose name is to be taken literally, and Lilli's Ballroom in Vienna is a barrier-free dance studio that provides access for dancing couples with visual impairments.

Learning to dance has many benefits for the blind, most importantly increased confidence. An encouraging "Dare to Dance!" is the motto of all such events because the hurdle to dancing tends to be very high for the blind. Poliana, a female ballet dancer from Brazil, confirms that when she says, "Blind people like me are sometimes afraid even just to take a step because we can't see anything. Dancing helps us to take that step. And I feel great when I do. Dancing liberates me."

Besides, dancing enlarges a blind person's repertoire of movements and trains their sense of balance. But most importantly, they profit from the amazing joy that comes with moving to music. If they engage in partner dances, the sighted dancer tends to lead, as did Artem Chigvintsev, Umstead's partner on *Dancing With the Stars*.

Tango particularly appeals to blind and visually impaired dancers because it is an intuitive form of dancing where a follower can trustingly submit to a leader. But it doesn't have to be partner dance. For more than twenty-five years, the Royal Opera House in London has offered so-called Monday Moves classes, in which blind and visually impaired dancers can study ballet. "Your body and your mind simply feel better afterward because you've done things you thought you could never do!" one participant says, describing her experience.

Similarly, the Fernanda Bianchini Ballet Association is a dance company in Brazil, near São Paolo, that has opened a dance school for blind children. For more than twenty years, blind and visually impaired children have studied ballet at the school and developed their repertoire of movements through dance. It is almost unbelievable how much life confidence these children have gained from their dance lessons. And it is most impressive to watch the school's dancers perform; after all, they have learned their art solely through touch and verbal instructions. They dance to the music with amazing precision and elegance—to check it out, watch the impressive documentary *Looking at the Stars*. "This is not only about dancing," says Fernanda Bianchini, the school's principal. "I'd love to leave all students with a piece of advice for life: don't ever stop fighting for your dreams."

In her bachelor's thesis, the Austrian dance teacher Lena Pirklhuber focused on the psychomotor development of blind and severely visually impaired children between the ages of six and ten. She argues that it is mainly their lack of experience with movement, caused by the visual impairment, that is detrimental to their motor development. Blind and visually impaired children often move carefully and hesitantly because they can't appraise situations or see dangers. But because movement is an important foundation for all perceptive faculties, any lack of opportunities for perception or action can impede a child's optimal development. Dancing involves the interplay of tactile-kinesthetic (touch and movement), auditory (hearing), and spatial (awareness of space) perception. Accessible to people who are totally or partially blind, dancing enables the integration of movement and expression in a holistic experience. This is why children with visual impairments can vastly benefit from dancing.

In Korea, a charitable organization that teaches dancing to the blind is called "Danceable Helen Keller" after the famous blind and deaf teacher. The organization mainly teaches a special meditation dance. Its goal is to help people develop better situational awareness and better control of their breathing, like a special mindfulness training.

Other opportunities exist to connect dance and the visually impaired. In Denver, Colorado, for example, the "Radio Dances" program has for twenty years described dances in such a way that blind listeners, too, can imagine what the performances look like. Many blind people love feeling so close to the dance stage in this "narrated" way.

Dance organizations for the blind also invite people without visual impairments to experience what it is like for blind people to go through their daily lives. For example, people are blindfolded to help them experience how dance is among the most intuitive forms of movement and very much an option even for those who cannot see.

Once you think about it, dancing while blind is actually not strange at all. After all, there are often moments in a dance when we close our eyes and blindly surrender to the music or our partner's movements.

But what is it like if we cannot hear the music? If it's not the sense of sight we lack but the sense of hearing? Can we still learn to dance? A scene in the movie *Beyond Silence* shows deaf children lying on the floor in order to feel music in their bodies. "She only loves music when it is loud. When the floor vibrates under her feet, she forgets she's deaf," German musician Herbert Grönemeyer sings.

Most likely you have experienced feeling the bass of a song: when your teenaged daughter cranks up the sound, when the stuff in your glove compartment is rattling to the radio, or when your whole body vibrates at a live concert. People can feel music. Deaf people can be even better at it because their other senses are more alert. They feel the beat and the rhythm and can dance to them. Also, acoustic signals can be translated into visual images so one can "see" the music. Deaf people are often quite visually oriented; whether lip-reading or watching sign language, they watch intently what the other person is saying. Most deaf people communicate expertly through movement and have learned to express themselves through movement—and, as we have seen, dance is another way for them to express themselves nonverbally. "For me as a deaf person, dance is the visual expression of what moves me inside, of what I feel. Whether you can hear or not is irrelevant in this regard. It's about making the inner movement visible on the outside of my body," said Doris Geist, a deaf sign-language instructor and dance trainer, in an interview. Greek researchers at the Aristotle University of Thessaloniki found that the fitness measurements of deaf people had markedly improved after a twelve-week dance training program, and in 2017, Rubianne Ligório de Lima at the Hospital de Cruz Vermelha Brasileira in Brazil reported definite improvements in deaf children's sense of balance after six months of practicing the fight dance capoeira.

The Nikita Dance Crew in Munich is the only dance group of its kind in Germany. It has dancers who can hear and also dancers who are deaf or hard of hearing. Their program is diverse and inspiring: sign language meets hip-hop and palpable beats. During training sessions, the woofer is placed directly on the floor so everyone can feel the beat. As Kassandra Wedel, the deaf dancer, choreographer, and founder of the group, puts it: "Music isn't just about hearing, you first have to feel it in your heart." The children's book *Through Sophie's Eyes* by the British writer Catherine Gibson tells the story of a young deaf girl who is determined to become a dancer. And as these examples show, it is by no means impossible!

The positive effect of dance goes further: children who have partially regained their hearing skills thanks to cochlear implants need to practice this new skill by attentively listening to specific sequences of notes, to different volumes, sounds, et cetera. Their brain must learn to decode the new sensory input and strengthen the corresponding neural connections. Dance can help in this process. A study by Tara Vongpaisal and her colleagues at MacEwan University in Edmonton, Alberta, showed in 2016 that cochlear-implant patients who listen to sounds while simultaneously doing dance movements progress much faster than patients who practice listening to sounds only. When we perceive sounds not only with our ears but also through the movements of our whole body, the learning process appears to speed up.

At the 2008 Paralympics in Beijing, blind and deaf dancers shared the stage and together danced the "Buddha With a Thousand Hands." The hearing-impaired dancers were given visual instructions, and the blind dancers followed the movements of the hearing-impaired dancers. The choreography worked without a glitch and was utterly beautiful.

Night has fallen, and we head back to the hotel. A cover band is playing at the bar, and we go straight to the dance floor to join in for "Dancing Queen" and "It's Raining Men." The combination of music and movement does the trick when it comes to our mood—certainly better than any drug! After three songs, any thoughts of the lost file in Germany or the upcoming lecture have disappeared. Even Julia's accident and Dong's life crisis seem far away—at least for this brief moment!

7

YOUNG AND OLD– DANCING AT ANY AGE

On with the dance! Let joy be unconfined.
LORD BYRON

Today a colleague has stunned us with numbers: more than 85 billion nerve cells with trillions of synapses form a unique pattern in a human brain. This network of nerves develops in childhood, takes years to mature, and is never fully complete.

"My father is a perfect example of this," Julia says with a laugh. "After he finished school, he took further training to become a telecommunications technician and then, at the age of forty, went back to school to study dentistry. Out of love for my mom he also taught himself Danish. The saying that you can't teach an old dog new tricks has always driven him crazy."

At long last, that saying has been scientifically disproven: we can learn new things no matter how old we are.

"And dance into advanced old age, too!" Dong adds.

To keep dancing as we age—what a wonderful dream. Legendary prima ballerina Maya Plisetskaya still performed the dying swan in pointe shoes at seventy-eight years of age; and despite being in her sixties, the singer Madonna keeps surprising us by doing the splits and multiple pirouettes in her shows. The Australian dancer and choreographer Eileen Kramer, born in 1914, most recently worked on a ballet that premiered in late 2017. Kramer is the face of the Arts Health Institute in Sydney, an institution whose goal is to use art, music, and dance to bring happiness and a better quality of life to older people. Each of these ladies looks at least twenty years younger than they actually are...

Legendary among swing dancers is the 1941 film *Hellzapoppin'*, in which Frankie Manning performs an incredibly fast, dynamic, and crazy swing dance with his friends. (Manning, together with George Snowden, is credited for having come up with the Lindy Hop.) One of the film's female dancers is Norma Miller, the "Queen of Swing," who was born in 1919 and who, before her death in 2019, was still an active swing dancer—a diva in a sparkling top and painted fingernails. In Harlem, where the Lindy Hop was created and danced with wild passion in the 1930s, a number of swing dancers from that first generation are still actively dancing today. Most are in their eighties or even nineties. I was lucky enough to see Miller dance at a festival to celebrate the one-hundredth anniversary of Frankie Manning's birth. Almost two thousand dancers from more than fifty countries gathered in New York for this one-of-a-kind festival. I had the privilege of dancing there with two legendary older female dancers of the swing scene: Barbara Billups and Sugar Sullivan, often referred to as "Barbara & Sugar." What an honor that was! Full of vitality, they laughed, joked, and even flirted with me.

Amazing, isn't it? Heartwarming and fascinating too. But what's that you're thinking? You aren't a world-famous star? So what? Dancing can still be a great addition to your life, especially if you're older, and even if you have mobility issues. Consider the example offered by the German Association for Senior Citizen Dancing. Located in Bremen, the association was founded in 1977 with the explicit goal of getting older people onto the dance floor. It creates dance routines that seniors can enjoy into advanced old age, with or without a partner; it also publishes dance instructions and CDs and offers annual professional development opportunities to support dance teachers in their work with seniors. The idea is catching on. As Eliane Gomes da Silva Borges at the University Castelo Branco in Brazil has been able to show, seniors' homes are increasingly deploying ballroom dance programs to help seniors regain autonomy and mobility.

Today's seniors' homes offer "chair dancing" or even "walker dancing" as part of their fitness programs, and instructions for such dances have become available on CD. In fact, special seated dances set to the "golden oldies" that seniors love have been developed and are now used in movement therapy. Many of today's seniors used to dance a lot when they were younger. Parties, weddings, and dinner dances provided abundant opportunities to hoof it and boogie. More than a few grandpas love to reminisce how they'd never miss an opportunity to dance with grandma—just to make sure no one else got the idea of approaching their sweetheart.

And let's not forget the positive impact of the music itself, so essential to dance. Music gives older people much joie de vivre and often brings back lovely memories. In her book *The Art of Grace: On Moving Well Through Life*, the American journalist, dance critic, and dance enthusiast Sarah Kaufman argues for adding more grace and beautiful movement to our lives. Grace can let all of us find more joy and new purpose.

Dance also affords many opportunities for social interaction/contact, which becomes especially important as we age and our social network shrinks. A person who dances has a social life. And if the partner with whom they have spent a lifetime of dancing is suddenly gone, dancing in a group may still make them feel good.

Many studies have demonstrated the importance of mental as well as physical exercise for staying healthy and fit in old age. As we have shown in the previous chapters, dancing supports both mental and physical health. But we now also have numerous studies specifically focused on dancing in old age. In 2015, researchers at the University of Hawaii undertook a meta-analysis of a large body of studies that explored the health benefits of dance. The analysis drew on participants between the ages of fifty-two and eighty-seven, and results showed that not only did the participants' flexibility improve significantly, but their strength and stamina also increased. In addition, they showed measurable improvements in their sense of balance and cognitive abilities. All these positive effects, by the way, were completely independent of what dance style was practiced.

Which brings us to another interesting discovery: researchers have found that dance styles we often (wrongly!) associate with being for younger dancers only, such as salsa, greatly benefit older dancers' sense of balance, strength, and stamina after just a few weeks of training. Besides, salsa dancing appears to notably minimize the risk of falling, as Urs Granacher and his colleagues at Friedrich Schiller University of Jena found in 2012.

Movement difficulties are a big problem in old age. Falls often cause fractures and subsequent complications. José Marmeleira at the University of Évora in Portugal was able to show that a dance program for seniors markedly improved their body awareness. Regular dance training teaches people to better appraise where their body, arms, and legs are at any given moment—and this awareness makes them feel more secure. A Japanese study by Ryosuke Shigematsu of the Foundation for Aging and Health has similarly shown that the risk of falling was significantly reduced after several weeks of dance training. After the program, the participants were better at standing on one leg than before, walked longer distances, and coped better with an obstacle course around a series of cones. Further studies have been able to show marked reductions in the risk of falling in seniors when they attend regular dance classes—for example, the studies by Olivier A. Coubard and his colleagues at the CNS-Fed in France, who had seniors take modern dance classes; or by Harvey Wallmann and Patricia Albert from the University of Nevada, whose enthusiastic seniors happily participated in jazz dance training (though without leaps).

Last but not least, an exciting study indicates that dancing can even help with incontinence. Older women who have had several pregnancies and births, in particular, tend to suffer from incontinence when muscle strength declines due to aging. Doctors often recommend special pelvic-floor exercises, but in 2017 So-Young An at Mokwon University in South Korea was able to prove that belly dancing has similar benefits. After twelve weeks of specific belly-dance training, the women reported significant gains, with measurable improvements in their pelvic-floor muscle strength.

> In Korea, there are special discos for seniors. Eighty-year-old men go there to dance five times a week—and dismiss fifty-year-olds as "too young." One typical Monday evening when curiosity took me to one of these discos, more than two hundred seniors were dancing. And did they ever tear it up! Some even take classes to learn cool moves that they can later show off on the dance floor.

DANCING TO PROTECT AGAINST MEMORY LOSS: DEMENTIA

Dementia is the worst nightmare of old age. According to the World Health Organization, nearly 50 million people around the world are affected, and so far, we have no medication to prevent it. Brazilian oncologist Drauzio Varella summed up the situation in 2010: "In today's world, five times more time and money are being invested in drugs for male virility and silicone implants for women than in the cure of Alzheimer's disease. In a few years we'll have many old women with large breasts and old men with stable erections, but none of them will remember what they are good for."

The word *dementia* is actually an umbrella term for disease patterns that involve impaired cognitive functions such as thinking, memory, and spatial orientation. People suffering from dementia are significantly hampered in their daily lives because they can no longer manage on their own. In the

United States, about 5.7 million people live with dementia, 60 to 70 percent of whom have Alzheimer's disease. By 2050, it's expected that 13.8 million Americans will be coping with some form of dementia. The risk grows with age because the nerve cells' structural connections slowly wither as we get older. The brain's so-called white matter shrinks, which results in reduced cognitive functions and slower processing of information.

In 2003, two separate teams of researchers published studies about dance and its impact on aging. Over a period of five years, the researchers had visited entire apartment blocks in the Bronx and had, armed with clipboards, interviewed residents between the ages of seventy-five and eighty about their hobbies. Do you do crossword puzzles? Play tennis? Play chess or cards? Dance? The researchers categorized the activities as either cognitive or physical hobbies. Neuropsychologists deployed tests to assess the participants' memory and cognitive flexibility. Their analyses left no doubt: only dancing offers some protection against dementia. An additional study confirmed their findings: compared to reading, crossword puzzles, card games, or playing a musical instrument, it was dancing that most significantly—by 76 percent—reduced the risk of developing dementia.

Agnieszka Burzynska of Colorado State University and her team wanted to find out whether a training program can arrest, or perhaps even reverse, the progression of dementia. They placed study participants into several groups. The first group was asked to go for a one-hour brisk walk three times a week and to change their diet; the second group was asked to take part in stretching exercises and balance training three times a week; and the third group was told to attend dance lessons. The dancers also met three times a week and mostly practiced complicated country-dance choreographies that involved passing from one partner to the next while dancing in lines or a square. Six months later, the brain's white matter had diminished in both the first and second groups, but not in the group of dancers. On the contrary, their brain substance had improved: it had grown denser in the area relevant for processing speed and memory. Several other studies confirm a similar effect. Dancing counteracts memory loss.

In fitness centers, exercise programs commonly prescribed for seniors often consist of monotonous, repetitive exercises, but these are insufficient to maintain brain fitness as we age. To maintain the brain's neuroplasticity and hence the formation of new neural connections, the brain needs complex tasks that require concentration. Notger Müller, researcher at both the German Center for Neurodegenerative Diseases and at the University Hospital Magdeburg, and his colleagues conducted a study in which research participants went through specialized dance training that kept them learning new steps and moves. The control group did fitness and strength training. Both groups consisted exclusively of seniors, with an average age of sixty-eight years. A range of tests, including cognition assessment, blood analyses, and brain scans, were done both at the beginning and at the end of the study. The results were baffling: over the course of six months, attention and flexibility significantly improved in the dancers. In both groups, brain scans showed structural changes and the strengthening of neural networks as a result of moving and of learning new movement sequences. In the control group, this new network connectivity was located predominantly in the areas responsible for motor skills and vision. In the group that had danced, the areas relevant for attention, memory, and complex movements had grown: complex dance training with new steps and unfamiliar sequences of movements appears to be more effective for the production of new nerve cells than simple fitness or strength training.

Dancing also has another fantastic side effect, as Azucena Gurmán-García and her colleagues at the North East London Foundation Trust were able to discover in 2013 in a seniors' home: all dancers felt better afterward. And so did their caregivers. Dancing increases well-being at all ages!

As previously mentioned, music has positive effects on dancers generally and on seniors specifically. It plays an even more important role in the context of dementia diseases because compared to other parts of the brain, long-term musical memory remains functional for an astoundingly long time, as researchers at the Max Planck Institute for Cognition and Neuroscience in Leipzig, the University of Amsterdam, and the University of Caen have shown.

An important characteristic of long-term musical memory is its close connection to our emotions—hence the so-called Play-It-Again-Sam effect. In the movie *Casablanca*, Ilsa (Ingrid Bergman) asks the pianist Sam (Dooley Wilson) to play, over and over again, the song that reminds her of her beloved. The stable connection of emotional and musical memory allows people to resurrect past feelings and can trigger veritable floods of memories. In music therapy, this effect is deliberately deployed with dementia patients to bring back joie de vivre and activate positive memories.

People suffering from dementia may barely recognize their partner but remember the melody of the song they first danced to together. The melodies of old songs often stay with us long after we have forgotten the lyrics.

PARKINSON'S DISEASE

"Do you think that one day you'll stop dancing?"

In the bar, after dinner, we can't stop talking about aging. It makes Julia think of her accident and the long period of time in which she did not dance.

"Maybe if I were to get sick. Or were in pain. Or got too frail…"

She sounds doubtful, and for good reason. Studies show that these are exactly the times when dancing is the best thing a person can do for themselves. "It's dancing or nothing!"

· · · · · · · · · · · · ·

My former neighbor Irma is eighty-four years old and for the past three years has lived in a seniors' home. She moved there when she felt she was no longer coping well with living alone in her own home. By then she had fallen three times, including once on her face. What had most shocked her was the fact that her body made no attempt to brace against the fall. She had simply collapsed "like a wet rag." The frightening diagnosis was Parkinson's. Irma was so afraid of falling again that she gave up her independence and moved into the seniors' home. When I visited her a few weeks later, I was quite surprised. She seemed much happier and more confident. The home's therapists had recommended she attend a dance class for Parkinson's patients. What at first sounded like a bad joke—"I'm no longer steady on my feet, so how could I possibly dance?"—has since become an indispensable part of Irma's daily life. Each time Irma dances, a small miracle unfolds: her stiff muscles and joints loosen, she feels the music, and the movements seem to follow by themselves.

Morbus Parkinson is one of the most common degenerative diseases of the nervous system. In 1817, the British medical doctor James Parkinson was the first to describe its typical symptoms. Parkinson's progresses slowly and mainly affects certain regions of the brain. The death of nerve cells in these regions results in insufficient production of the neurotransmitter dopamine, which in turn disrupts the transmission of nerve impulses between brain and muscles. Patients increasingly lose control over their movements and experience uncontrollable tremors, paralysis, and—frequently—falls, which often lead to further complications, particularly with seniors. The incidence of the disease increases with advancing age. More than 10 million people around the world live with Parkinson's—approximately 60,000 new cases are diagnosed in the United States each year and another 6,600 in Canada. No cure exists, but medication can help to mitigate many of the typical symptoms.

Studies show that dancing, and especially tango dancing, can lead to noticeable improvements in the mobility of Parkinson's patients. In one study, American neuroscientists Madeleine Hackney and Gammon Earhart had fifty-eight patients with a mild form of Parkinson's undergo dance training. For one hour twice a week over a period of thirteen weeks, participants danced waltz or tango or did exercises unrelated to dance. Compared with the non-dancers, the dancers showed improvements in their motor skills, such as walking distance, step size, and balance, with the tango dancers outperforming the waltz dancers. Additional studies confirmed the effectiveness of tango dancing. This may be due to the fact that tango dancers almost always have to momentarily stand on one leg and often have to pause and start up again. Starting up again is an especially big challenge for Parkinson's patients, but once they are in the swing of it, things tend to go fairly well. Characteristic tango figures are good exercises for starting up.

In a comparative study, researcher Marie McNeely and her colleagues discovered it was not only mobility skills that improved in Parkinson's patients but also mood, cognition, and general quality of life. The same conclusions were reached by Erin Foster and her colleagues at Washington University School of Medicine: after Parkinson's patients had attended a twelve-month tango training program, they had an improved sense of balance and were also more motivated—compared to the non-dancing control group—to keep up their familiar activities and daily routines, such as going out for meals or socializing with friends.

And that's not the end of tango dancing's benefits: in 2013, Kathleen McKee and Madeleine Hackney of the Emory School of Medicine were able to show that after three months of ninety minutes twice-weekly tango dancing, Parkinson's patients experienced a reduction of symptoms as well as important cognitive improvements in comparison to a control group that was only given health education. The dancers had notably better spatial awareness—an awareness often negatively impacted by the disease.

These findings speak for themselves and, fortunately, are increasingly translated into practice. In 2001, for example, the Brooklyn Parkinson Group began, together with choreographer Mark Morris, to offer dance classes for patients and their families. The classes combined elements of ballet, jazz dance, and ballroom and aimed not simply to make participants move but, more importantly, to have them move together and let them regain trust in their bodies. This particular group concept is now being taught in many countries as "Dance for PD," or Dance for Parkinson's Disease.

DANCING FOR CHILDREN

Today we arrive at the bar a bit earlier than usual. On the dance floor, a DJ is fiddling with their equipment, and we're curious as to what style of music will be played tonight to get us dancing. Suddenly, an oversized bunny with the hotel logo on its chest jumps out from behind the scenery. We look at each other, dumbfounded. The speakers start playing "Veo Veo," and in no time at all the dance floor fills up. Everybody dances enthusiastically to the popular Spanish children's song, right here in Greece. Or at least everybody shorter than four feet three inches.

The benefits of dancing in the last stage of life apply to dancing in life's early stages as well. Young children in particular have an intense urge to move. They constantly run, jump, skip, wriggle and crawl, balance and climb. Moving so much allows them to constantly learn new things. They get to know their environment, probe unknown territory, and discover themselves, their body, and their body's limits. Earlier in this book, we argued that moving to music is innate and simply a given for children, and we've shown how dance is good for our motivation and our brain. Supporting a child's natural need to move, then, is worth the effort. We can do it even with very young children and almost in passing. Try, for example, to combine the dreaded task of tidying up with dancing. If Lego blocks are put away to great music and the "tidying-up dance," the whole business is a lot more fun—both for children and their parents.

In educational theory, it is becoming more widely accepted that children who dance regularly have better learning skills and are more even-tempered during learning experiences in general. The internet has a lot of videos that show how to explain facts with dance. Search "math dance" or "dance by the numbers" to check it out.

Elizabeth Spelke at Harvard University found that years of dancing promotes better spatial awareness in school-age children. Children who are passionate about dancing did noticeably better in geometry tests than children who had only danced for a very short period or not at all.

Teachers around the world try to take advantage of the beneficial impact of dancing on the brain and its neural flexibility. At Hayward Middle School in the United States, for example, mathematics is taught exclusively in combination with music and dance, and student grades in math have steadily improved since the introduction of this model.

Dance and movement can help to make other subject areas more easily understood, too, and no one who has watched such presentations will forget the information delivered this way. The science slam by Mai Thi-Nguyen Kim from Germany, for example, has become legendary: she uses hip-hop to explain chemistry and chemical processes.

The prestigious *Science* magazine is also emphasizing the connection between knowledge and dance. Each year, it organizes a unique competition: "Dance Your PhD." Doctoral students explain their dissertations and experiments nonverbally through dance. It's unbelievable how many ways there are to express and interpret knowledge through dance movements.

When content is linked to rhythm and movement, it isn't only learning that improves. Rhythmic games and uncomplicated choreographies let children develop better body awareness. Like seniors, children develop a sense of safety when they get to know their body through dance movements; they develop a better assessment of their body and its movement in space than children who don't dance. Dance has children draw on more multimodal skills than other types of sport: they have to simultaneously listen, watch, move, anticipate other dancers' moves, and adjust their own, all of which requires flexibility, coordination, strength, and stamina.

And that is important for their development. The World Health Organization recommends at least one hour of physical activity per day for children and youths, and yet reports that 81 percent of adolescents aged eleven to seventeen were insufficiently active in 2010. Moving to music comes naturally, even for children who may struggle with being overweight or are too shy to assert themselves on the soccer field. Dance is not about the number of goals or the "higher, faster, farther" demands that children are often subjected to from an early age, but about movement for its own sake.

Dance education can start as early as age three. Numerous approaches use imaginative play to teach children how to move to music. And starting young has benefits. During preschool and kindergarten, boys typically have not yet developed an aversion to dance. On the contrary, they enjoy any kind of movement and find moving to music particularly fun. It is only once they have internalized gender stereotypes from their environment, generally during elementary school, that boys start to avoid dancing. Even if boys later throw in the towel, their earlier dancing will never have been in vain.

The German soccer player Wolfgang Dremmler, who played on the national team and for a long time served as talent scout for the Bayern Munich soccer club, recommended dance to soccer-crazed boys—exactly because dancing fosters so many skills during a child's development. Good dance teachers can motivate boys by choosing a theme and letting them play robots dancing on Mars, for example, or enact a fight dance like capoeira.

Some dances are so "in" that no persuasion is needed. In the 1990s, Michael Jackson's moonwalk conquered the schoolyards, while in 2019 the floss dance from the video game *Fortnite* got boys all over the world moving. It's even been performed by football players after a touchdown.

As we have described, children intuitively express themselves by moving to music. Children's dance has room for much improvisation and creativity. Children skip, stamp, or fly like butterflies. They twirl around the room, spinning around their own axis, till they drop, literally. This way, children deepen their nonverbal means of expressing themselves and come to terms with events in their daily lives.

Dance provides a space for children to adopt different roles. Shy Lena, for example, can become an evil monster dragon, and Glen, a prima ballerina on tiptoes. It's just a moment, but in this moment, during their dance, they can be and do anything.

By the time they start elementary school, children often know very clearly how they like to dance—in a tutu to classical music, like Taylor Swift in a video clip, or with the dramatic moves of break dancing. Letting children see the range of dance options is worth it. Why not attend a ballet with your child or take the time to watch a hip-hop crew performing on the street?

If you take your child to dance class, make sure the teacher is competent. Having been trained as a teacher surely matters more than having been a performer on a big stage. Above all else, working with children demands skill, patience, and empathy.

Watching children dance is an uplifting experience, and we're almost sad when the children's turn on the floor is over. Afterward, the band starts up again, and the older children who haven't been sent to bed remain on the dance floor for the first songs.

"Look at him!" Dong points at a boy who seems to be about eight years old and is trying to teach his grandma the floss dance. She will need a bit more practice, but she's already doing quite well. After all, neuroplasticity stays with us into advanced old age!

8

DANCE DOES EVEN MORE—LET'S LAUGH, CRY, AND DANCE

Sometimes you get lucky and find a soul that dances to the same beat as you do.
ANONYMOUS

This is our last evening, and the conference participants are sad about having to say goodbye. We've had so many discussions and have learned so much from each other's research. Above all, we've seen how intertwined neuroscience and dance really are.

Our group has become a team. We plan to stay in touch and continue to exchange ideas. Our "salsa couple" aren't the only ones who feel there's more to come!

THIS IS ALSO ABOUT SEX

In the Bible, Salome dances, and when she is offered a reward, demands the head of John the Baptist. Her story has inspired countless artists: Oscar Wilde wrote a stage play; Richard Strauss composed an opera; and many painters, such as Picasso or Munch, have chosen to represent her in their work. In the 1953 movie *Salome*, Rita Hayworth performs the "Dance of the Seven Veils." Whether on screen or stage, a woman who dances tends to symbolize eroticism, seduction, and, ultimately, danger. Women's dancing supposedly causes men to lose their minds and self-control. Many traditional stories tell of the "fatal dance"; Mata Hari's dancing, for example, is said to have ruined several military officers. Does dance really have such enormous powers of seduction?

Do you know the 1980 dance movie *Fame*? If so, do you recall the dance-class rebel Leroy, who wanted to protest against the establishment and especially ballet, which he considered rigid, boring, and somewhat superficial?

Check out this scene: One day, Leroy arrives five minutes late to the ballet studio. In one hand, he carries a towel, in the other, a heavy boom box. Beads of sweat glisten on his dark skin; it's obvious he's already been training hard. Twenty-one pairs of eyes follow as he—gum-chewing and provocatively slow—struts across the room and takes a free spot at the barre. The stern ballet teacher with the glasses fixes a reprimanding gaze on him, her mouth a thin line. He defiantly returns her gaze, puts down his boom box, and turns it on. A throbbing beat shatters the silence. The well-behaved girls stand lined up on the right side of the room, sneaking glances at the teacher, unsure of how to react. But then their eyes settle on Leroy again. He swings his hips as he moves to the center of the room. Adoringly, the girls gaze upon the beauty of his neck and shoulders and the litheness with which his body pauses along with the music before bursting into wild dance. The teacher raises an eyebrow, but her mouth is no longer quite so tight. Driven by the music, Leroy's movements grow more expansive. Stepping forcefully to the beat of the music, he gyrates, his hands suggestively accompanying the movement of his hips, his upper body lightly undulating. Back and forth... Head to toe, the man is one erotic dance move.

Leroy's dance is meant to provoke ballet's rigid establishment, and for that he gives it a sexual energy no one can ignore. The pianist stares at Leroy; all the ballerinas at the barre stand gaping; and even the furious teacher is unable to completely resist the erotic charisma of his dance.

Procreation is a basic need of any species; without it, the species would go extinct. The human brain has evolved with a mechanism that ensures we understand the eroticism of a movement—an ancient mechanism that comes into play no matter how civilized we think we are. This, too, is about body language: our brain knows what we would feel if we were to move the way Leroy was moving.

Clever choreographers have used the reverse strategy and deployed very "proper" dance genres such as ballet to communicate erotic messages.

When rehearsing the ballet *Marguerite and Armand* by Frederick Ashton, world-famous ballerina Margot Fonteyn is said to have commented that the rehearsals contained "a passion more real than life itself." A fantasy—and a ballet that audiences love to this day. We can watch it through two different lenses: as beautiful, chaste ballet or as a metaphor for a sexual act. The interpretation of the piece is literally in the eyes of the beholder.

A dance move can thus be multilayered. It can make a political statement, send an erotic message, or simply be beautiful. Sometimes a dance gets interpreted in ways that were never intended. Did you know, for example, that Middle Eastern belly dances were originally performed by Egyptian men in baggy harem pants? In the early twentieth century, enterprising coffeehouse owners in Cairo noticed that their well-heeled European patrons were particularly pleased if it was women who danced. And so, the shrewd businessmen sent women onto the stage instead of men, according to some anthropological accounts.

Generally, our perception of dance is very much shaped by the culture from which we come and by the dance forms and other means of expression we have encountered. What we know today as pole dancing and consider highly erotic is actually a complex artistic dance form from India with deep spiritual meaning, completely misrepresented in the West. In its original form, it was certainly not performed by lightly clad or nude dancers. But often, accounts by European travelers depicted these culturally rich dances as erotic or linked to prostitution, simply because they did not understand them through their Western looking glass. A similar fate befell some Iranian and many African dances.

Before we pass judgment on erotic dancers, we should remember that what we might *perceive* as erotic in our culture may not be intended that way. Their dance may simply be an expression of certain emotions or a cultural practice with which we're unfamiliar. We should try not to fall into the same trap as people back in colonial times.

Nevertheless, some researchers—particularly those who study evolution—are exploring the seductive angle of dance. It could be said that they are on a quest for the ultimate, universally erotic dance move.

A team of researchers working with British professor Nick Neave at Northumbria University wanted to identify any movement by dancers of the opposite sex to which our brain responds particularly intensely—that is, those movements that strike us as being especially attractive. The researchers filmed men and women as they danced in a purpose-built studio in which every move could be recorded. The research participants wore small sensors on their bodies that transmitted data to a computer about the speed, rotation, and size of each movement. This allowed the researchers to collect video clips of recreational dancers of both genders and to create a database about movements of each body part. The researchers then used this database to generate computer-animated, faceless, dancing avatars so that another group of men and women who were subsequently asked to judge the dancers' attractiveness would genuinely assess the movements and not the person's looks or charm.

What men especially liked about female dancers was a big hip swing, asymmetrical movements of the thighs, and the ability to move both arms independently of each other. Women found greater variation and size of movements by male dancers' necks and upper bodies irresistible. The more expansive the men's bends and turns, the more the women liked them. Surprisingly, the speed of their right-knee movements was a favorite, too.

Their right knee? In Leroy's country of birth, Colombia, people dance salsa. They start dancing practically before they can walk, and they continue to dance throughout life to the rhythm of Latin music. Even though in salsa dancing both sexes move their hips and upper bodies extensively, it is really the ladies' hip swing and the gentlemen's upper-body gyration that epitomize the dance. And when the men lead the women, the men will often do a sidestep with their right leg, which they tend to cushion with a smooth flexion of the knee joint; simultaneously, the women will raise one arm above their head, asymmetrically to the other arm. Funnily enough, the results of the study do confirm, in a way, the sex appeal of this style of dancing.

What do these results mean for you? Ladies, if you're looking to attract the male gaze you'll do well to make wide circles with your hips. Choose moves that allow your thighs and arms to move asymmetrically to the beat. And for gentlemen who might be looking for a lady: prepare for an onrush when you've found the right combination of neck, shoulder, and right-knee movements. Let your upper body undulate. Bend it far sideways and move your head sideways, too. Don't forget your right knee: the more twists and sidesteps, the better.

It can also be useful to throw your dance partner somewhat off-balance. But only if you know what you're doing and are sure you'll be able to catch her... In 1974, psychologists Donald Dutton and Arthur Aron did an experiment in which male research participants either had to walk across a wobbly suspension bridge above rocks and cliffs or across a safe, sturdy bridge. Halfway across, the men in both groups encountered the same woman, who would ask them for an interview and give them her phone number for any follow-up questions. When asked by another researcher about the woman they had just encountered, the men who had met her on the suspension bridge found her more attractive and called her notably more often than the men who had met her on the solid bridge. They had erroneously interpreted their fear—in reality caused by the wobbly bridge—as excitement about the unfamiliar woman. In psychology, we speak of the "misattribution of arousal." Someone who gets a dance partner to release a bit of adrenalin into their blood stream has a better chance of going home with that partner's phone number.

I know a dancer who loves to dance with beginners. He purposefully throws them a bit off-balance and then uses his skill to catch them with a smile. It sets off butterflies in the ladies' stomach, in the same way the suspension bridge did for the men in the study. One after the other, the beginners fall for him.

Some evolutionary biologists believe that human dance originated as a courtship ritual. Elvis certainly used the power of his swinging hips to his advantage, and no one who has watched *Dirty Dancing* can deny that dancing is often about "the birds and the bees." Tango, salsa, or merengue—dancing can be a highly erotic experience. In fact, when we dance, our bodies experience similar biological processes as when we have sex: endorphins, testosterone, and oxytocin are released and trigger feelings of happiness and exhaustion, just like after we've made love.

But in reality, eroticism and attractiveness are only possible side-effects of dance, not its raison d'être. The misconception that dance is solely about sex has caused some countries to ban dance. That's a pity and a sign of ignorance. For dance can do and be about so much more, and it's unfair to reduce this miracle cure to a single aspect.

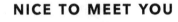

NICE TO MEET YOU

So far, we've spent every evening here at the bar and have not once gone to a dance club in town. It's a pity, really.

"Let's check online to see what's available."

Dong pulls out his smartphone, and we google "milonga," "Lindy Hop," and local clubs. We are lucky: this very evening, and quite close to our hotel, a beach party is taking place.

"Should we check it out?"

A rhetorical question, by the sound of it.

· · · · · · · · · · · · ·

At the University of Konstanz in southern Germany, where I used to work as a tutor in the international office, exchange students from more than eighty countries would arrive at the beginning of each term. Most of them would be randomly assigned to the dorms, where every night over the first weeks, wild parties would take place. Nothing would bring these students together as quickly as dancing with each other. Michael Jackson, Prince, and Madonna worked many wonders. By the time songs like the "Macarena" were played, everybody had thrown themselves into the action, and it no longer mattered whether a student had come from Spain, Colombia, the United States, or the other end of Germany.

Especially when we are in a foreign country, dance offers a wonderful opportunity to connect. Most travelers browse their guidebooks to find a cozy café or a trendy bar where they might have a chance to interact with people who live in that particular place. Unfortunately, they rarely meet as many people as they'd imagined they would, because the patrons at coffee shops or bars probably arrived with their friends, are engaged in conversation, or, if they happen to have come on their own, are flipping through a magazine or reading a book. It takes a lot of courage to insert yourself into a conversation between people you don't know or to simply approach someone new. It's even more difficult when we don't speak the local language well enough or at all.

But there's a straightforward way to meet new people even in foreign places—and it comes with guaranteed success. Whether we dance salsa, swing, or Argentine tango, we just have to find out when and where that particular style is being danced as "a social" (that's what social dancing events are often called). Generally, that kind of information is available online or in the events calendars that are often available at newsstands in bigger cities. You go there, and you dance. It sounds easy, but the astonishing truth is that it is easy. Often, you don't even have to speak to a potential partner because a look or a gesture may be all that's needed to get them onto the floor. People may chat, but they don't have to. Not speaking is never embarrassing when we have something to do and a step sequence and music to focus on. And maybe, just maybe, a conversation will come about at the bar after a wild dance number. We already have one thing in common, after all: we're dancing the same style of dance and may well listen to the same kind of music. If enough people do this often, and in many places, something like a community is formed—and it's quite possible that we'll run into someone in Vancouver that we last danced with in Seattle.

For years I had not seen my childhood friend Shin. But I had heard she was now living in Seoul. One day, out of the blue, Shin called me.

"I need you, urgently. Can you come to my place next Monday night?"

"Are you okay? Is everything all right?"

"Yes, it's all fine. Don't worry. It's nothing bad. But I do need you. And it's urgent."

"Sure, I'll help. But can't you tell me what it's about?"

She didn't reveal anything, just gave me a date and an address in Seoul. It was almost like a thriller!

A week later I stood in front of a building in Shinsa, which is in Seoul's Gangnam district (yes, the one in the song). A staircase led down to the basement, and I imagined any number of scenarios. Was Shin in trouble? Threatened by evil gangsters? Had she borrowed money? Why did she urgently need my help? What had led her, and now me, to this building?

I walked down the hall until I came to a soundproof metal door. Without any idea of what to expect, I opened the door.

Wild swing music was pulsing; sweaty swing dancers looked at me, and Shin rushed up and welcomed me with a happy smile on her flushed face: "We badly need men to dance with! Thanks for coming!"

I don't know exactly how it happened, but by the end of that evening I had become a swing dancer. I got to meet a lot of new people, many of whom I am still friends with today.

What works in foreign places can also work at home, where it's even more worth our while to make new friends. But many find it especially difficult to go out on their own in their hometown; they're afraid of being labeled a lonely freak. If we take a closer look at a club's dance floor, though, we'll probably notice a few people in the crowd who dance by themselves. Besides, being by oneself doesn't have to mean being lonely.

Dance events like "socials" and "milongas" are often more suitable for getting to meet someone than anonymous, loud clubs. When it comes right down to it, we have nothing to lose and everything to gain. If the event turns out to be boring or unpleasant, just leave. But beware... don't mistake such evenings for dating events. Surveys show that many men hope to find a new partner this way. Of course, this does occasionally happen, with new dance partners falling head over heels in love the very first evening they dance together. But it's worth noting that women usually don't go to a dance in order to meet men. They want to dance, feel free, and express their feelings. That's what Aniko Maraz and her colleagues from Eötvös Loránd University in Budapest found in 2015. They surveyed 457 dancers about their reasons for dancing. The ladies gave the following reasons in this order of priority: stay or get fit, improve their mood, get into a trance, become more confident, and escape their daily routines. The gentlemen gave just two important reasons: intimacy and flirting.

Could the gentlemen perhaps learn from the ladies?

When we go dancing on our own, our attention is not focused on the group or our friends but on ourselves in the here and now. We concentrate on the dance and not on the unfamiliar people who share the dance floor with us. It's quite possible we only get to know the truly interesting people once we pay attention to them.

Dancing is also a good opportunity for groups whose members don't yet know each other. A workshop to learn new steps makes everyone confront a problem they will ultimately solve collectively. As we saw earlier, this type of activity lets group members bond and makes their brains fire in synchrony. These are excellent conditions for strengthening a team. No wonder some agencies offer dance events as incentives and team-building seminars. Resistance to change can be an especially big issue in workplace teams, where team members are often too much in their heads. Movement makes the emotional side of coworkers accessible. And there will be practice, cursing, and laughter.

Afterward, you feel much more closely bonded with the people you have danced with than those you have not. Your brain has posted a "memo to self" to remind you that these people are special because you and your body have shared something important with them: a dance.

ENJOYING DANCE FROM A DISTANCE

The beach party is in full swing, and every young man and woman from the local village seems to have shown up. The music plays with a constant, unchanging beat. The dance floor is made of straw mats, stapled together. The two of us definitely raise the average age. Fascinated, we take our drinks to a couple of beach chairs and watch the spectacle.

"Watching others dance also changes your brain," Julia says with a laugh.

Australian researcher Catherine Stevens and her team at the University of Western Sydney found in a series of studies that we can experience intense emotional arousal even if we only watch others dance. Our body shares the thrill and excitement of a dancer we watch leap across the stage. Frank Pollick, too, did studies on this phenomenon, including work with his colleagues Corinne Jola and Seon Hee Jang. In 2011 and 2012, they published research that showed that the motion perception in the brains of people who like to watch dancing can be as trained as if they were performing the dance movements themselves. Can we learn to dance through observation?

The painter Josep Coll Bardolet was extraordinarily successful in representing dance on canvas. Most of his life he lived and painted in the village Valldemossa on the Balearic island Majorca. To the bell of Chopin's Charterhouse ringing in the evening and the reverberating echoes of the mountain goats' bells welcoming the night in their own way, Bardolet painted scenes of daily life in Majorca, among them the traditional boot dance, or *ball de bot*, so typical of the island. Village festivals on the island last deep into the night, and to this day the Majorcans dance the *ball de bot* at these festivals. However, one has to leave the touristy areas to see the real deal, which has little in common with the folkloric razzmatazz that is performed for the tourists and depicts the Majorcans as medieval hillbillies. The *ball de bot* is a bit similar to the tarantella, the Italian anti-stress dance. It is wild and bursts with energy, and its dancers leap and spin. These are the particular movements on which Coll Bardolet's paintings focus. Many artists have tried to capture dance on their canvas, but even Edgar Degas's world-famous ballerinas are rather static still lifes. Not so Coll Bardolet's dancers. You're practically pulled into his paintings, as if you were leaning into the breeze and were invited into a wild dance. "What I now draw in five minutes," the painter once said, "took a lifetime of observation."

In fact, when we observe movement in a picture—a movement frozen in time—our brain is active with the same mirroring and movement processes that would be occurring if we were watching the action live. Professor David Freedberg is an art historian from South Africa who now teaches at Columbia University in New York. During his time as director of the Warburg Institute for Art History in London, he introduced an exciting new line of interdisciplinary research. Within the walls of this renowned institution, there now is a lab that draws on both art history and neuroscience, today led by Professor Manos Tsakiris. Some of the first interdisciplinary assessments that Freedberg and his colleagues undertook analyzed how our brain responds to movement in paintings and drawings. If, for example, we contemplate Goya's *Désastres de la guerre*—a series of paintings that depict the horrors of Spain's occupation under Napoleon—a brain scan shows activation of the same regions that fire when we experience real pain.

Neuroscientist Beatrice de Gelder, an expert on body language and emotions, has also examined what happens in our brain when we gaze at movements on canvas. When we see a jump or a spin, be it in a painting or in reality, our brain gives us some of the same butterflies we would feel if we were actually doing the jump or spin ourselves. And while we may feel the butterflies in our stomach and not in our head, it is the brain's mirror mechanism that sends them to the stomach in the first place.

What still images can do, real-life movement can do better. Our mirror neurons fire when we watch someone dance. In 2012, Corinne Jola and her colleagues were able to show that regular attendance at dance performances activates the mirror system in our brain. Although the spectators sit calmly in their seats, the activity unfolding in their heads is similar to that in the heads of the dancers on the stage. However, the spectators' brains mirror not so much the precise fine motor skills of professional dancers as the processes that would occur in the brains of inexperienced dancers doing those movements.

Consider the following scene: a ballerina enters the stage. Lights are dimmed. A single spotlight follows her delicate figure as she glides through space in complete silence. Her steps are inaudible. Then, very softly and from afar, a violin is heard. The sound grows, becomes louder, and is utterly beautiful. When the audience, entranced by the sound, inhales like a single organism, the ballerina interrupts her path. She doesn't pause, though; instead, gently following the violin's rising sound, she begins to raise her right leg. The audience is spellbound by the ballerina's movement and fixated on her beautifully shaped leg, which draws an arc as it rises, seemingly light as a feather. The ballerina adds to the tension created by her leg's arc by lifting her right arm in a similarly graceful arc. Up, higher and higher up, following each rising note of the violin. When the violin reaches its highest note, the ballerina's right leg forms an almost vertical line from the floor. The audience holds its breath. All of a sudden, with a thundering drum roll, the entire orchestra starts up, and the ballerina exits the stage with a massive split jump. The audience starts to breathe again.

Movement, and especially *dance* movement, fascinates the human eye and calls on our brain and heart. The reason for this is rooted deep in our evolutionary history. Our ancestors were able to survive only if they could perceive and assess any movement in their environment. What has moved? Is danger at hand? Do I need to ready myself for battle? Also, only those individuals who knew how to correctly interpret movements by members of the opposite sex had a chance to procreate. This is a good reason why our visual perception is so highly developed and why we react to movement quickly, intuitively, and intensely. But not every movement grabs our attention—that would be too exhausting. Specialized neurons for movement ensure that only "important" movements claim our full attention. These are movements that signal danger, procreation, food, and safety: horizontal movements that alerted our ancestors that something was approaching or running away; expansive, unexpected, or angular movements that generally meant danger; and gentle, rounded movements that signaled protection and safety. The evolutionary scientist Irenäus Eibl-Eibesfeldt further described body poses that can be found all over the world in rock paintings and may have been expressions of fertility. Consider, for example, certain splayed poses in Tahitian and Australian rock art. Interestingly, we find all of these movements and poses in today's dance styles. Dance represents, and plays with, humankind's essential movements, and our brain responds.

The human brain is fond of large, impressive movements. If they don't indicate danger but simply show how well someone can do something, we love to watch because we like to admire. Our brain understands the challenge of a movement. Watching dancers perform gives our brain great pleasure. To be filled with amazement can be so wonderful.

However, beauty has only recently become a topic of interest for neuroscientists. Only in the mid-1990s did neuroscience discover the arts. When and why do we perceive something as beautiful? What happens in our brain when we encounter art? It is thanks to pioneers such as Anjan Chatterjee and Semir Zeki that a new branch of brain research has developed: neuroaesthetics. It examines how we perceive art, what emotions art triggers and how, and which brain functions are responsible for our creativity.

The brain reacts particularly strongly to certain shapes and movements. For example, in 2006 Moshe Bar and Maital Neta studied participants who lay in an MRI machine and were shown numerous different objects in random order, including sofas, teddy bears, flowers, and abstract shapes. Each object was represented twice: one variant with rounded corners and another one with sharp corners and edges. The participants identified how much they liked each image, and they clearly preferred the rounded variants. But what was even more interesting was this: when they were shown the variants with sharp edges and corners, they generally judged them negatively, and their brains activated those areas responsible for dealing with threats. Researchers explain this phenomenon as stemming from the danger that sharp objects pose to the body and the need, therefore, for us to regard them with cautious apprehension.

That same year, Joel Aronoff arrived at similar conclusions in a study on painting. Images with angular shapes were judged more negatively than images with rounded shapes. Aronoff and his team then expanded their research to include ballet. He designed an ingenious test, which drew on video recordings of five classic ballet performances. Participants with no prior ballet experience were asked to watch the videos and mark any segment they interpreted as showing either positive or negative dance roles, such as the black swan and the white swan, Odile and Odette, in *Swan Lake*. Another group of participants was asked to categorize these dance sequences according to round or angular movements. Negative characters such as the black swan danced longer sequences with angular movements, while positive characters danced longer sequences with round movements.

Whether a movement is round or angular is not the only aspect our brain is interested in. The brain loves to be impressed. Even though movements in classical ballet delineate the natural limits of the human body, dance movements have changed over time. In a 2009 study, Elena Daprati and Patrick Haggard at University College London analyzed pictures of performances of the ballets *Sleeping Beauty*, *Snow White*, and *Giselle* from the archive of the Royal Opera House, London. They measured the angle of the dancers' legs in 1962, 1979, 1996, and 2003 and found that over the years the performance of the movements had changed considerably even though the choreography had remained the same: today, more extreme poses are more important than ever before. While the 1960s ballerinas turned out and extended their working leg behind their body for the so-called arabesque at about ninety degrees, ballerinas today raise the leg by almost 180 degrees, and impressively straight lines have become imperative.

Our working group thought these results were very exciting, and in 2016 we conducted a further study on that same question. Together with Professors Antoni Gomila, Frank Pollick, and Anna Lambrechts, we asked research participants to assess six-second-long ballet video clips. Their assessment was unambiguous: round was seen as positive, angular as negative, and the more extreme the poses and the higher the raised leg, the more they were liked by the research participants.

The anticipation spectators experience when a dancer raises their leg is like a deliciously exciting thrill. The leg rises and rises... and higher still? In chapter 1, we mentioned that anticipation is particularly pleasing to our brain. Fittingly, one of the biggest moments in *Swan Lake* is when the black swan comes on stage and from a standing position does thirty-two fouetté turns. Like a music box ballerina, the dancer turns and turns, and each of her turns is a perfect piece of art, similar to but not identical to the previous one. The hypnotized audience cannot take their eyes off her; they count along as the suspense and their admiration keep growing: one... two... three... twenty-two! Twenty-three!... Thirty! Thirty-one! Thirty-two! Their anticipation has been rewarded.

We experience this scene as pleasurable because our brain loves repetition and because we feel a certain amount of suspense and thrill in spite of the repetitions: "Will she pull off another fouetté without falling?"

Many dance styles also use brief pauses for effect. The ballerina may stand *en pointe* just before her final position... and stand... and stand... before she'll finally ease out of her gravity-defying pose to move into her final position. In tango dancing, too, there are moments that appear to be interruptions—but they are merely a pause before the storm re-erupts a moment later. In the so-called *parada* the leader places his foot against his partner's foot and makes her stop, in a way tripping her but not making her stumble. He leads her close to his foot, and it is up to her to step over it when and how she wants to. Tango often plays with this little thrill: Where are we heading next?

Several studies have shown that anything we are unable to immediately anticipate makes us wonder what will follow. If we don't get what we expect right away, we either become angry and frustrated like a toddler at the supermarket checkout, or—provided the arc of suspense is designed properly—we become curious. We are unable to look away and excitedly await what comes next. We want more! However, often more is expected of the audience than mere admiration of impressive movements, especially in modern dance. Dancers tell us a story that may not always be easy to comprehend—or tell no story at all, which can be deliciously confusing and interesting too!

Imagine four dancers, each waiting in their corner of the wings for their entrance on stage. The lights go down, and the first dancer moves. Ready?

He carries a torch, stomps fearsomely, and leaps abruptly a few times into the air. He must be carrying a rattle because he is creating a horrible racket. Then the second dancer follows. Accompanied by a musician playing the flute, he quickly bends his knees and straightens up, down and up, and down and up again... If we pay close attention, we notice that he actually looks like a rooster strutting across the farmyard. Continuing with the rooster theme, the third dancer enters, a male dancer entirely covered in feathers. There doesn't seem to have been enough money for a proper costume, though, because these feathers have clearly seen better days! Dull and crooked, they stick to his body. The fourth dancer seems not of this world. His moves suggest he is endlessly swallowing something, and the exaggerated burps that accompany his gulps are grotesque. To top it off, he steals some popcorn and a water bottle from a member of the audience...

What was all *that* about? And why have we spent so much money on the tickets!? The audience at a dance show sometimes wonders, "What on earth is the choreographer trying to tell us with this?"

Do you remember analyzing poetry in high school? Wasn't it mind-boggling what the teacher led us to discover in those poems? All the talk about meter, the speaker, and metaphors—a garden stood for the Garden of Eden, a wall for a problem, a thick branch for a strong life (or a penis, depending on what grade you were in). In the beginning, you just sat there, confused; it all seemed to be gibberish. But once you learned to read the images, it stopped being so difficult. Everywhere in the world, students learn how to decode literary imagery through reading the literature of their country. A similar teaching method would be useful for the understanding of certain dances.

"Say, is it shadows that flit unclogg'd by the load of the body? / Say, is it Elves that weave fairy-rings under the moon?" Even in Friedrich von Schiller's poem "The Dance" question marks appear. It's not your fault, then, if you don't understand the message of a dance performance right away. To understand dances that have been shaped by their culture of origin, we need more information than just the language of movements that our body speaks right from birth. According to French philosopher Michel Foucault, the body is a medium that communicates messages. But the meaning of these messages is only accessible to people familiar with their social, historical, and cultural contexts. And because dance has always been closely tied to the culture and religion of individual tribes and peoples, no single, unified history of dance exists.

There is, however, an academic discipline devoted to the analysis of the meaning of dance. In the early 1980s, dance scholar Professor Judith Lynne Hanna set out to develop a system for the description and classification of individual movements and their meaning. After talking to dancers all over the world, she came up with a system that can be used for dance movements from different cultures. For Hanna, dance is a complex system that consists of small units, the movements, that when taken in context add up to words, the dance vocabulary. These words can be combined to form complete sentences and hence messages. Depending on the social context in which the dance is performed, the sentences can communicate different meanings. The same, of course, is true of spoken language. For example, if we call a muddy path "slippery," the word means something entirely different than if we call someone a "slippery fellow." In a similar way, dance movements can take on different meanings, and we have to be very careful when interpreting dances with which we are unfamiliar. In modern dance, especially, interpretation can be very difficult.

I once attended a ballet where the choreographer explained the individual poses. One dancer stood with his legs spread and his head on the floor in front of him. With his buttocks aimed at the audience, he winked at us from between his legs. Two other dancers had adopted similarly twisted positions. The choreographer took the mic and began to explain this first scene, pointing at the buttock dancer doing his half headstand.

"As you can see, this person clearly has many questions," he said. What? How would I know? If someone in my social circle has questions, they rarely point their buttocks at me. Most of the audience seemed to be just as confused as I was, so the choreographer added, "When I see a scene like this, I wait for it to evolve. I let it affect me, and possibly an answer to one of my

Dance Does Even More—Let's Laugh, Cry, and Dance

questions will emerge. Or not. Which may also be an answer. Namely that this question wasn't mine at all, but one of the dancer's questions, and that while I, too, may have questions, this one wasn't one of them. Perhaps I'll find the answer to my own question in his next movement." *What?* Did that explanation help you? It certainly didn't help us.

But at times, this confusion seems to be exactly the point. A dance doesn't always need to mean something. Maybe the dancer just kicks their leg up higher than you have ever seen. Modern dance doesn't necessarily have to be understood; it can simply provoke questions, provide food for thought, and take you to new places in your mind. And the more experience we have watching dance, the better we can perceive the many levels of meaning in its movements. Trust us. Practice makes perfect. Even the practice of watching.

Of course, you can ask a dancer after the show what a particular dance meant, but much more important is what the dance meant *to you*—for its meaning to you will most certainly reflect what you are dealing with, what *you* feel, or what *you* need.

A while ago, a good friend of mine who also happens to be a dancer invited me to an international dance festival, where I went to several modern-dance performances by international choreographers and dancers. I remember particularly well a show from Israel. The dancers formed a circle and all danced perfectly to the same beat. But every so often a dancer tried to break out, escape from the circle, and dance something different. Each time, the crowd of other dancers pushed them back into the circle and forced them to dance the same way as everyone else.

Remarkably, when the dancers who had tried to escape had been forced to become part of the group again, they acted in unison with the group and prevented other dancers from breaking out. Everyone was both: an individual pulling away and a member of a captive crowd. The play continued until, one after another, the dancers in the circle slowed down, crouched down, and eventually stopped moving altogether. This dance, although performed entirely without words, told a story. It was a story I could understand and identify with as easily as if it had been my own. And maybe it was my own: at the time, I went to a Korean school where we all had to wear the same uniform and learn by heart the same material and where individual thought was not encouraged. Any attempt to break away from the crowd was doomed. Even though the dance originated in a completely different culture, it touched me, and I intuitively understood it. Seeing this show was incredibly fascinating and one of the key experiences that brought me to dancing.

Santiago Ramón y Cajal, the famous pioneer of neuroscience, is reported to have said, "Any man could be the sculptor of his own brain." You already know that dance experiences, like any other experiences, create neural structures in the brain. Emily Cross and Louise Kirsch, former dance students and now neuroscientists, specialize in research on dance and the brain and have published several studies on the topic. For one of the studies, they put their research participants into one of three groups. With the help of a video game console, one group was to learn several dance sequences set to different pieces of music. Another group was merely shown the music videos but not asked to do the dances themselves. And the third group only listened to the music. Before and after the training sessions, Cross and Kirsch had their participants assess a multitude of dance sequences, including those they had practiced or watched, respectively. Only the group that had actively learned the steps liked the dance sequences better the second time. We prefer the dances we have tried ourselves because our body memory helps us to interpret these movements. We "understand" them better, and what we understand better, we like better.

For many dancers it doesn't matter what the audience sees. Instead, what matters to them is to express authentically what they are concerned with themselves. We sense their strong emotions and authenticity and let our own brains react.

Watching others dance does something to us, which is why dance as an art form tends to affect our lives. If you want to understand a dance performance, keep your eyes open for movements you recognize and that are easy to interpret. If a dancer stamps and jumps, he is probably full of angry energy, but if he glides across the stage, he likely wants to express lightness. Listen to the music; it tells us much about the mood.

You may even have a program for the show that explains some of the scenes and metaphors. But most important is to allow yourself to feel what is going on inside of you. There is no right or wrong. The expression of feelings is universal.

INTEGRATION AND "ACCEP-*DANCE*"

The beach party takes some getting used to. We are given a once-over, and some of the young male dancers put on a big show in front of Julia. It may well be an advantage that we don't understand Greek! We interpret their adolescent carryings-on as resentment—until, that is, a particularly formidable, well-muscled young man approaches us and with a nod and a debonair smile invites us to the dance floor. It's time to let go of a few prejudices!

Michaela DePrince was not always Michaela DePrince. At the beginning of her life her name was Mabinty Bangura. Like thousands of other Muslim children orphaned in Sierra Leone's civil war in the 1990s, she had ended up in an orphanage after the murder of her parents. One day, the wind blew a magazine cover across the orphanage's dusty yard.

When Mabinty plucked the cover from the dirt, she saw a ballet dancer *en pointe*. The image took her breath away, not because she had never seen ballet but because she had never seen a white person. This "amazing creature," as Mabinty called the figure on the cover, looked very happy. In Mabinty's mind, this beautiful being offered a glimmer of hope amidst her life's bleak desolation. Mabinty was the least-loved child in the orphanage. She was bullied and abused and was given less food than the others, the most faded clothes, the worst of everything. Mabinty has vitiligo, a non-infectious skin disease that causes the loss of pigmentation in blotches. Black skin turns white, and because of these special white sparkles in her skin Mabinty had come to be called the "devil's child" in Sierra Leone.

Mabinty kept the picture of the white ballerina. Hiding it in her clothes, she took it with her wherever she went, and often pulled it out. Whenever she looked at it, she was able to push the horrors of war from her mind. She wanted to be like the being on the cover: happy. When she was four, a Jewish family in the United States adopted her, and Mabinty became Michaela. This, now, was *her* happiness! She had a home and felt loved and safe. And she was given a chance to dance. Dance gave her strength. And the memory of the mysterious beauty in the photo accompanied her like a talisman on her journey. It stood by her when, at eight years of age, she was not allowed to dance Clara in the *Nutcracker* because "America's not ready for a black Clara," and when at age nine, she heard that a producer was unwilling to invest in her because as a black girl she would "soon get wide hips and big tits." But nothing could stop the slender dancer Michaela DePrince. She trained with iron discipline and never took her eyes off her goal of becoming a radiant, happy dancer. At fourteen, she won the Youth America Grand Prix, a prestigious ballet competition from which the up-and-coming dancers of the next generation emerge. Supported by her adoptive mother, Michaela conquered the world of dance. She graduated from the renowned Jacqueline Kennedy Onassis School at the American Ballet Theatre and today is one of the few black prima ballerinas who dance the great solo roles. These are often roles that tell life-or-death stories, and there are not many dancers her age that can tell these stories as convincingly as Michaela DePrince. She has also written a book, *Taking Flight*, and performed in a movie. And she's never lost track of her goal: she has danced, across boundaries and borders.

Unfortunately, participating in our society's social and cultural life still depends—too much—on one's background. People who, for a variety of reasons, are on the margins of their community have a difficult time finding a place for themselves in society. Studies show that integration tends to depend on cultural education. Open access to a society's cultural life is not only important for an individual's personal development but also strengthens the community and hence society at large. Everyone benefits if people, regardless of their background and social status, engage with one another through cultural activity. In that way, mutual understanding of cultural and social differences can grow.

Many marginalized people find comfort in dance. Vogue, or voguing, is a dance style that developed in 1960s Harlem. Its moves are inspired by fashion models and their poses on the catwalk and were initially performed mainly by gay or transgender African American males. Rejected by mainstream America, these dancers created so-called houses that, for many of them, provided substitute families. They mimicked the glamorous world to which they had no access in real life.

In recent decades we have come to recognize that dance can create community and, with it, acceptance. Dancing allows people to be different *within* a community. What matters is personal development and, ultimately, motivation and working toward one's goals, because any dance project requires tenacity and discipline. An excellent example of how this works comes from a unique and inspiring project out of Germany. In February 2013, the Berlin Philharmonic Orchestra, under its principal conductor Sir Simon Rattle, started its first great educational project. Led by choreographer and dance teacher Royston Maldoom, two hundred and fifty children and teenagers from Berlin schools, representing twenty-five different countries with their respective cultural and social environments, danced Stravinsky's ballet *The Rite of Spring*. The choreographer had had some international experience with such projects, having earlier performed Stravinsky's piece with street children in Ethiopia and young offenders in England. For most of the youths in Berlin, though, the project marked the first time they'd encountered classical music and dance. "You can change your life in a dance class," was the motto Maldoom used as he challenged his young performers to push themselves to their limits in order to make them aware of their enormous potential. During the project, many of the young dancers went through emotional highs and lows and were tormented by doubts, but ultimately, they outdid themselves. The whole process was recorded, and these recordings became the foundation for the wonderful 2004 documentary film *Rhythm Is It!*, which vividly shows the students' development and has thrilled more than half a million viewers around the world.

Today, many such projects exist. Not all are as large and as effectively geared toward the media, but the fact that dancing can significantly contribute to students' education and hence to their motivation and integration is becoming increasingly obvious. Just consider what world-famous ballroom dancer Pierre Dulaine and his team were able to accomplish. In 2013, Dulaine—who was born in Jaffa, Palestine, in 1944—embarked on an ambitious project in the city of his birth, now part of Israel. A documentary film team accompanied the team as they set up "dancing classrooms" with Palestinian and Israeli Jewish students. The children were paired up in mixed teams to learn to dance, and to respect each other. At first, it seemed an impossible task—initially, the children wouldn't even touch each other's hands—but over the ten weeks that the program ran, it became a model of integration, friendship, and mutual respect. The resulting movie, *Dancing in Jaffa*, is recommended viewing for anyone wanting to see the power of dance to unite even the most divided parties.

Other studies have looked at hip-hop and break dance to see whether dance can in particular improve the academic performance and health of socially disadvantaged youths. In 2017, Justin Bonner and his colleagues at Morgan State University in Baltimore, Maryland, showed that youths who dance hip-hop have better memory skills and higher cognitive flexibility. These same youths were also better at reading positive emotions in photos than their contemporaries who were without hip-hop experience—they were better able to empathize. The researchers see in their findings a promising approach to using hip-hop in order to reach young people who otherwise show little interest in learning the standard curriculum.

This is exactly the approach that, for example, the nonprofit organization Hip Hop Public Health pursues. Its aim is to use music and dance to raise young people's awareness of health issues and to motivate them to adopt a healthier lifestyle. That this strategy can really work was shown in 2017 by Cendrine Robinson and her colleagues at the University of Maryland who analyzed studies on hip-hop-based interventions and health: apparently, dance culture furthers young people's health awareness. For example, in the context of one hip-hop project, young people were provided with information about strokes, which they evidently not only absorbed but also passed on to their parents.

Whether in a tough inner-city neighborhood, a refugee camp, or a school, many small steps can be taken through dancing to further integration.

 A few years ago, I took part in a production of a new version of Schiller's classic stage play *The Robbers*. We were a motley crew: classical musicians, jazz musicians, DJs, and professional dancers with ballet and modern-dance experience. The classical musicians and ballet dancers represented the play's noble court and aristocracy, the jazz musicians and DJs the play's ordinary people. The robbers were played by break dance kids who usually performed in the streets. The director had approached them to see if they would like to participate in an open-air project, and they agreed. Although our backgrounds included different musical and dance styles and our social environments also differed widely, we became a team. As soon as we were on the stage and began to dance, we were able to forget everything else; it was truly astounding. Such amazing projects can overcome social difficulties and interpersonal conflicts. They really are small miracles!

Royston Maldoom, the choreographer of *Rhythm Is It!*, has gone on to supervise other fascinating projects. All over the world he produces and directs "community dance projects" designed to enrich people's lives through dance and to build bridges between cultures. In Germany, he has also worked with refugees. For people who have lost their homeland and feel uprooted, dance is a wonderful tool for bringing together their different worlds and expressing their emotions in the process.

"When I dance, I feel back at home," says Arif Mohammed, a Syrian doctor who now lives in Dresden, Germany. He is a passionate tango dancer and has discovered in his new home city a lively tango community. When he closes his eyes and lets the music lead him and his partner, the music and movements take him back to Aleppo. How is it that the Syrian Arif can feel so at home in a German city during evenings of tango dancing? Are the movements of his legs the cause? Is it the music? Or is it his brain?

You know by now that people, objects, and certain customs or actions—as well as certain sounds or music—trigger memories in our brains. Music and dance, in particular, are often connected to memories of daily life, of beautiful festivities, and of big family celebrations. They are markers in our brain that let us feel "at home." Music and dance give us feelings of security and comfort because they are familiar and connected to positive emotions. When you dance a dance from your childhood, you are at home, whether it is the chicken dance of a four-year-old Canadian or, in this case, the Argentine tango music of an uprooted Syrian. Maldoom once said, "Dance is a very special art form. It is a physical, emotional, spiritual, cognitive, and social activity." This is why dance offers us the opportunity to change our life and our personality for the better, to redefine ourselves in a new community, and to let us belong even though we may be different.

In spite of its awkward start, the beach party is a fitting end to our time on the Greek island. Dancing among these unfamiliar young people, we experience firsthand how beautifully dance unites us, regardless of our place of origin, language, or age. Dance is a language spoken and understood everywhere, an international, indeed universal, language that lets us understand dancers from cultures all over the world. On the dance floor, we all are and feel equal.

LET'S DANCE

Every year, April 29 is celebrated as International Dance Day. The designation dates back to 1982, when it was created by the Dance Committee of the International Theatre Institute, a body affiliated with the United Nations Educational, Scientific and Cultural Organization (UNESCO). The date was chosen in recognition of the birthdate of Jean-Georges Noverre, a great reformer of ballet.

The intention behind the day is to allow for the celebration of dance as a universal language. From Australia's Aboriginal people to the Inuit of the Far North—regardless of culture, religion, politics, or ethnicity, dance is a global phenomenon.

And it is a truly ancient phenomenon. Our ancestors most likely intuited all the benefits of this art of moving long before the science of neurotransmitters, mirror neurons, and aerobic fitness came along. Dancing trains our motor skills, self-perception, and memory, and it enhances our freedom, our creativity, our emotions, and our community. It strengthens our cardiovascular system and our immune system, improves our posture, and keeps us nimble and flexible into old age. Dancing improves our mood and confidence and, coincidentally, provides a great workout that makes us lose weight and tighten up. Finally, and most important, dancing goes straight to the brain and strengthens the connections between nerve cells: we learn more easily and keep mentally fit. What other movement can do all that? Dance is a miracle drug. So why do so few people dance?

According to a recent public health survey in the United States, about 35 percent of girls and 8.4 percent of boys report dancing as part of their regular physical activity. Overall, the prevalence of dance participation among US adolescents was nearly 21 percent—leaving nearly 80 percent who don't dance at all.

"We should write a book about dancing," Dong says with conviction. We are in a cab on our way to the airport. Dong will fly to Berlin, Julia back to London. "Yes, we should!" In the cab, Julia starts jotting down notes on a napkin.

As we arrive at the airport, we learn that our cabdriver understands German and has been following our conversation. After eight days of scientific discussion and debate—eight days during which we've studied, exchanged ideas, and danced—a roughly sixty-year-old Greek cabdriver sums it up best: "Dancing is wonderful! People always say laughter is the best medicine, but really, it's dancing. And you get laughter as a bonus!"

It's time to conquer the dance floor—to, as David Bowie once suggested, put on our red shoes and dance away our troubles. Whether it's waltz or hip-hop, a quickstep or even Julia's pony farm dance, whether in a club, a dance class, or at home in your living room, just put on your favorite music and dance. And if you can't motivate yourself to do it alone, go ahead and sign up for a class. There are plenty of opportunities out there, even online! Dance for kids or ballet for seniors, contact improvisation or folk dance, headbanging or the blues—there are so many facets to dance that there's something for everyone. Fitness dance programs such as Zumba or Bokwa have also become more widely available. Try out different styles of dancing and take your time as you look for the one that's right for you. What works for a friend or spouse may not work for you. And once you have found the right style, don't expect everything to work out perfectly right away. Most dances aren't learned in one hot summer, like in *Dirty Dancing*, and not every dance has to be "The Time of My Life." Be aware of your own pace, your own rhythm. Michael Ende put it beautifully in his children's book *Momo*: "That music came from far away but rang out deep inside me." Find your own music deep inside you, and dance to it.

Dancing freely is something that every single one of us can do—as long as we can pluck up the courage. As for certain dance style steps... well, it's a big mistake to think you ought to know or be able to do something before you have learned it. You would never expect yourself to turn a somersault or to pole-vault without training. So why do it with dance? Everyone was a beginner once.

Just enjoy the movements. As we have seen, happiness, health, and mastery will follow on their own. Why? Because our brain and our body can't help but dance.

9

DANCE TEST: WHICH STYLE FITS ME?

....

You got to dance like nobody's watchin'.
SUSANNA CLARK

....

Six months have passed since the conference on Aegina, and we are meeting on the island Majorca to draft the book's final chapter: the dance test.

Together we drive into the mountains to the village Valldemossa where Coll Bardolet painted Majorcan dancers, where the Charterhouse reminds us of Chopin, and where wonderful coffee shops invite us to philosophize for hours on end. We talk, for example, about how to find the dance style that suits us best.

IT HAS TO FIT

"Every person has a talent, you just need to find the one that's for you," says the commentator in the dance film Billy Elliot after Billy is knocked out in the boxing ring. Billy's talent obviously doesn't lie with that particular sport...

But how do we find what's best for us? And when it comes to dance, how can we know which style is right for us? In this chapter, we'll explore a few dance styles in order to help you find some answers to these questions. Perhaps you'll feel inspired to try out one or more.

My friend Anna Lambrechts is a researcher, swing dancer, and swing-dance teacher all rolled into one. Together with her husband, Simon Selmon, she teaches at Swingdance Society UK. The society's courses and dance evenings take place in a building that is a business school by day and a dance palace by night. On Saturdays, swing and salsa dancers share the space. Salsa—fun and somewhat erotic—has its home "downstairs," in the dark catacombs, while swing—bright and cheery—resides on the first floor. Every so often this leads to bizarre encounters in the bathrooms that both dance communities share. Worlds collide! For a start, imagine the dancers' different outfits. Anna grins as she describes the looks the dancers give each other. *Really?* they seem to ask. *Why would anyone wear that!*

The best way to discover whether we like something is to ask ourselves if we want to do it again. This is pretty self-evident in other contexts. Does someone go for a second helping at the dinner table, for example, or do they reluctantly pick at their food and decline to have more?

If we like something, we want more of it. This is true of dancing too. Once you've found a style that is right for you, no effort will be too much, no journey too far to attend your next evening of dance. But if you have to badger yourself and your inner couch potato is pushing you back into the cushions with all its might—and you happily give in—well... then you haven't yet found your dance.

> It took me a long time to find "my" dance style. I always enjoyed watching ballet, especially *Swan Lake* and *The Nutcracker*, but ballet was not something I wanted to dance myself. As a teenager, I was shy and didn't have much courage when I dabbled in break dance, so that didn't work out so well. A little later, during my first term at university, I ventured into ballroom dancing, which didn't turn out too splendidly either. When I was an exchange student in the United States, my Latin American roommates tried to get me excited about salsa, with middling success. I was able to dance it, but it didn't exactly blow my mind. I much preferred to dance freely and solo to the funky beats of drum 'n' bass or deep house. That's when I was able to really let go–though it was less the dance itself that entranced me than the music and being with other people at the club.
>
> It took me years to realize that I'm a swing dancer, that all of it–the music, the dance movements, the community–is perfect for me. When I did, it felt as if I'd finally found my family. The swing dance community has become a home for me.

Of course, you can't constantly be running to dance schools, sports associations, or clubs to try out yet another number of dance styles. So, in order to find the right dance for you, the first question you can ask yourself is this: Do I like the music that this particular dance is danced to? If you can't stand Latin rhythms, a salsa class is probably not the best place for you. Go and surf YouTube; you'll find music and dance videos galore. And don't forget: "can't do that" is forbidden. You can learn it.

CHECKLIST: WHAT DO YOU NEED FOR A DANCE CLASS?

First things first: you don't need any big equipment to dance. Your body, your mind, a bit of openness, patience, and the willingness to experience something new are all that you need. Invite your inner child to come along, that part of you that feels joy but no embarrassment, because it will be a big help, especially in the beginning... Beyond that, here are a few more tips you may find helpful.

- Trust the findings of developmental psychology: not everybody is staring at you. Really. There's nothing for you to be embarrassed about.

- Wear shoes that fit you well, are comfortable, and are stylish. (Never step onto the dance floor wearing brand-new shoes, and don't wear shoes with plastic soles. Instead, wear leather soles on parquet and other wooden flooring, and on stone flooring, rubber soles that still allow you to slide. Always protect your knees.)

- Wear comfortable and stylish clothes (if you wear your pajamas or something that looks like them, don't be surprised if people seem a bit confused).

- Take an extra shirt or top to change into in case you sweat a lot. And even though we've written about the power of pheromones and that we shouldn't always cover up every person's individual scent, a good shower with soap and the use of deodorant is still advisable.

- If you still feel a bit unsure of yourself, here's a final tip from the pros: go to a different part of town or to another city where nobody knows you.

MAY I? DANCE!

BALLET

Having originated at the French court of the Sun King (Louis xiv) in the fifteenth century, classical ballet is now an art form that captures the imagination of people of all ages and levels of expertise (from lay audiences to professional dancers). People who dance ballet as a hobby mainly dance solo. During training, much emphasis is placed on elegance and body control. This dance style is most admired for its abstract aesthetics, the lightness of its movements, and the beautiful feeling that dancers derive from moving elegantly to classical music. Ballet consists of a distinct system of steps that carry French names such as "pirouette" or "arabesque" and are taught the same way almost everywhere. Dance classes will begin with exercises at the barre, followed by exercises in the center of the room, and will conclude with more expansive movements, such as leaps across the entire space. Pointe work is not necessarily a must and is not offered at all ballet schools.

Do you like to dance solo? Do you love classical music and dance movements where the focus is on the lines and shapes of the body in space? Do you prefer clear rules and set sequences of movements with little room for improvisation? Can you imagine dancing in front of a mirror?

If you've answered "yes" to these questions, you may enjoy attending a trial lesson. Ballet classes are available everywhere, and for every age.

ARGENTINE TANGO

Argentine tango emerged in the early twentieth century in the night clubs of Buenos Aires where European immigrants sought to soothe their loneliness and longings with dance. The core feeling of Argentine tango and its music is melancholy, though it can also fill you with exuberance and energy. Because of Argentine tango's cultural significance, it was inscribed in 2009 on the Representative List of the Intangible Cultural Heritage of Humanity by UNESCO. Argentine tango, or *tango Argentino*, refers to the original form of this dance and its music and is distinct from the strictly prescribed ballroom tango.

Argentine tango is elegant, authentic, and feels like a conversation between two bodies in one embrace. During a tango lesson, dancers pair up, stand in a circle, and practice their steps with their partner. Depending on the school, the lesson may include changing partners or not. In some schools, students only dance with their own partner. There is a manageable number of basic steps that are easy to learn. What's exciting about Argentine tango is that the different steps and turns can be combined in any way. Everything can be "said" with tango's basic vocabulary. Also, opportunities to dance the Argentine tango abound. In addition to the dance classes offered in just about any city, there are milongas (social dance events where Argentine tango is danced), tango marathons, and tango festivals. During milongas, the leader invites the follower with their gaze (called the "mirada")—so be careful who you look at! Tango songs are always played in a string of three or four. This is called a "tanda," and couples dance the whole tanda with the same partner. Does the melancholy and passion of Argentine tango music appeal to you? Do you like partner dancing? Do you enjoy leading or following? Can you imagine getting creative and improvising with the prescribed basic steps?

If this speaks to you, you should definitely pamper yourself with a trial lesson in Argentine tango.

BALLROOM AND LATIN AMERICAN DANCES

We address these dances together because they tend to be taught together and form the bulk of the World Dance Program that was developed in the 1960s by the World Dance Council, a body constituted of national dance teachers' associations. The program comprises the dances typically taught by dance schools.

Ballroom dances probably date back to the Renaissance, when they grew out of a mix of folk and court dances. In the West, they were the first respectable partner dances. Today's ballroom dances are the slow waltz, the Viennese waltz, ballroom tango, foxtrot, slowfox, and quickstep. Latin American dances, often simply referred to as "Latin dances," are the samba, cha-cha, rumba, paso doble, and jive. Although only the first three are truly from South America, the five dances are grouped together because their technical elements make them very similar. Social dances such as the bachata, merengue, or salsa don't belong in the category of "Latin dances" in the ballroom realm, even though they come from Latin America. In addition to the rumba and the cha-cha, North Americans dance the Spanish bolero, East Coast swing, and—instead of samba, jive, and paso doble—the mambo, which stems from South America too. Each dance style has its own basic steps, posture, and even distinct clothes and shoes. The dances can be performed on a stage (for entertainment or in competitions) but are also categories that hobby dancers specialize in. As a hobby dancer, you can earn badges and participate in competitions, which for many dancers adds motivation. In addition to the classes, dancers often have opportunities to attend balls or to dance with other students who have successfully completed the series of classes in their respective category. The offers are endless, as a quick online search will show.

Are you open and flexible about a whole range of dances? Do you like to follow prescribed step sequences with a dance partner? Would you enjoy working toward getting different "diplomas" for your dancing? If you want to be ready for the variety of musical styles you are likely to encounter at a dance event or in a club, ballroom and Latin are the perfect choice for you.

ZUMBA

Zumba is still a very young dance style. In the 1990s, the Colombian Alberto Pérez wondered, "Why should we dance Latin only as couples?" He took elements from the Latin dance styles and developed them into a sweat-inducing fitness dance that anyone can do alone or in a group. Every class consists of a mix of Latin dance movements and common fitness exercises, accompanied by stirring Latin music hits. Compared to straight fitness programs with music, Zumba has no uninterrupted, continuous beat; instead, its movements vary according to the songs played. People dance to the rhythm of the music, with the step combinations coming about almost intuitively. During a class, the dancers mostly form one large group and, facing a mirror, copy the teacher's movements. Sometimes they also study short step sequences that get developed further in subsequent training sessions. This fitness dance trains the dancers' coordination, strengthens their muscles, and is terrific for building stamina. You can burn up to four hundred calories in a single Zumba class! But regardless of one's fitness level or age, Zumba works for almost anybody, because the training intensity can be modified for each individual. Variants of normal Zumba are tailored to children, seniors, or people with physical limitations. There's something for everyone.

Do you like Latin rhythms? The feeling of burning off all your energy? Do you look at dancing from a fitness perspective? If you do, this rhythmic workout in front of a mirror is probably your thing. Check it out. Many fitness studios offer drop-in or beginners' classes.

SALSA AND CO.

Salsa, bachata, merengue, rueda Cubana, danzón, kizomba, zouk... The list of dances in this category is long. They developed in the twentieth century in Latin American countries where they have always been, and still are, part of daily life. Salsa music made it to North America in the early 1900s, and the dance style soon followed, reaching the height of its popularity in the 1970s. A lively salsa scene now exists, especially in bigger cities. There are salsa bars, and many clubs organize salsa evenings. Almost every dance school now offers salsa classes. Above everything else, salsa is fun, and its dominant vibe is pure joy with an erotic touch.

Each of these dances has a prescribed vocabulary of basic steps. Once you have mastered it, there are no limits as to how you can improvise with that vocabulary. Salsa movements have numbers ("80," "50," "60"), and fun names such as *dile que no* ("tell him no"). Teachers will teach these steps or short step sequences one after another. Dance partners change regularly, so you will quickly get to know many new people. You often dance quite close.

Do you love Latin music? Do you enjoy leading or being led? If you don't mind being close to a dance partner and love to move to hot rhythms, this is where you want to be. Pluck up the courage. There's sure to be a class somewhere close to you.

STREET DANCE
(HIP-HOP, BREAK DANCE, AND CO.)

The street dances include hip-hop, break dance, vogueing, and many more. These styles originated in the poorer neighborhoods of American cities in the 1970s. People met in the streets, tried out the latest moves, learned from and admired each other, and competed in "dance battles"—and still do. Confrontations or conflicts were resolved through dancing. To outsiders, these dance moves can look somewhat aggressive or provocative. The moves often follow the lyrics very closely, directly translating text into movement. Social bonds within the street-dance community are very strong.

Street dance is shaped by improvisation and draws on dance moves from all other dance styles. That way, break dancing mixes with hip-hop and sometimes even with modified classical dance movements. As with most group dance styles, lessons typically begin with warm-up exercises before studying shorter choreographies, which will then be expanded on in subsequent training sessions. Lessons usually end with some space for improvisation. Opportunities for street dancing are available literally in the streets, but also at so-called conventions. Some of the street-dance moves also lend themselves to free dancing in discos and clubs.

Anyone under the impression that hip-hop and break dancing are limited to youth is wrong. A quick online search of "hip hop for adults" will lead you to classes in your area.

Do you enjoy typical street dance music? Are you less into perfection and set choreographies and more into opportunities for expression and release? Do you like acrobatic moves in a dance, maybe even done in synchrony with a group? If you do, you may want to try out these styles. Many sports clubs and dance schools offer lessons.

MODERN DANCE AND CO.

Modern dance is a rather vague term to categorize some very different genres but is often thought to include contemporary dance and neoclassical ballet. What these styles have in common is that they use movement for the genuine, authentic expression of feelings and thoughts. When modern dance emerged at the beginning of the twentieth century, its pioneers regarded it as a reaction to the prescriptiveness of ballet, whose movements they deemed unnatural and artificial. Modern dance is thus often understood as anti-establishment. It doesn't matter whether the establishment is a nagging boss, a despotic state, or the constraints of ballet: in modern dance, freedom of expression is the prime principle.

Originally, modern dance took no distinct forms but primarily served to freely communicate emotional experiences. Over time, various techniques and characteristic moves did develop, but genuine expression has remained its ultimate objective. Movements in modern dance are at times as light and soaring as in classical ballet but can also be forceful, loud, and heavy. Refreshingly few rules exist, and the music is eclectic. Percussion and piano music are as common as rock, techno, or pop. There are even dances without any music at all, where the rhythm is the dancers' breathing. Lessons in modern dance are so widely available that there's something for every taste and age. Teachers will encourage students not to chase after abstract aesthetics but to really focus on themselves and their movements, their inner genuine expressivity. Mostly, classes will start with warm-up exercises and then offer students the opportunity to study short choreographies with the movement vocabulary of the particular style of the teacher.

Do you enjoy freely moving to music? Would you like to learn a dance style that doesn't constrain you and gives you ample opportunities to express your feelings? Are you in the mood for getting a bit closer to yourself through movement?

If you answered "yes" to these questions, give this style a try and let yourself gain self-experience through modern dance.

JAZZ DANCE

Jazz dance emerged in the early twentieth century in the United States and most likely has its roots in African American social and street dances. Since its early days, it has developed a distinct style and can be found almost anywhere in the world.

Jazz dance is both a stage and a hobby dance. Most of the dance schools that teach ballet and modern dance also offer jazz dance, which emphasizes body control. Similar to ballet classes, a class in jazz dance teaches specific techniques in order to precisely train posture and steps. A lesson consists of warm-up exercises, repetitions of dance routines, and the study of new, more complex choreographies. Jazz dance movements involve the entire body and include turns, floor exercises, stretches, and, often, leaps. People dance to the latest popular music. It's worth visiting a variety of dance schools or going to one of the many jazz dance workshops available worldwide to get inspired.

Would you like to dance solo—without a partner—but in a group? Learn choreographies to the latest music? Move through space with expansive movements and leaps? Do synchronous movements and group formations appeal to you? Could you maybe see yourself performing with a dance group?

If you answer these questions with "yes," you should try out jazz dance.

FOLK DANCES

There are as many folk dances as there are cultures in this world. Today they are often danced at local festivals. Originally, though, they were not intended as performance pieces but sought to strengthen the community and identity of the local culture. Why? People who dance together develop strong bonds. Therefore, if you move to a new place, it may be a good idea to attend a class on regional folk dances. You'll be part of the community in no time.

Folk dances usually have clearly prescribed step sequences. Some are actually very simple (as in the tarantella), while others are rather complex (as in flamenco). Some are partner dances, while others are danced solo or in a group. The latter is the case with the Majorcan boot dance, the Argentine chacarera, or the polka mazurka. For many folk dances, you will have the opportunity to try on traditional wear, or be invited to bring specific accessories such as scarves, little bells, or other metal or wooden objects that produce sounds during the dance.

Do you feel closely attached to your region and culture, or have you just moved to a new region and/or culture? Do you enjoy learning about other cultures? Would you like to meet new people through dance? Anyone who treasures traditions, local customs, and the feeling of togetherness might find a "happy home" in a folk-dance class. They are often offered by local clubs.

AFRICAN DANCE

Let's begin by acknowledging that using a single label to refer to the dances of an entire continent is, of course, a huge oversimplification. But we have decided on this approach because dealing with the abundance of African dances goes beyond the scope of this book.

Many of today's dance styles—such as swing, jazz, Argentine tango, and square dance—are said to have roots in African dance. In Africa, music, dance, and song are inextricably linked, and they shape people's cultures and views about their daily lives. Dance often marks a society's important events, as is the case with fertility dances, war dances, and initiation rites, and also features in many spiritual events. Typically passed on from one generation to the  next, they are important markers of cultural identity. Although different in style, many of the dances follow a typical pattern, so that a beginner can join in after little practice and "groove along." African dances are often very complex: whereas in the West a mainly unified movement of the dancer's entire body is desired, dancers of African styles often maintain several spatially and rhythmically distinct focal points for their body's movements. Sometimes shoulders, torso, arms, and legs follow different rhythms. Such dancing definitely requires control over one's body! The rhythm follows alternating tension and relaxation, and the movements often have an accent downward, "into the ground," which distinguishes African dances from Western dance styles where posture is mostly upright and meant to signal lightness, with an accent "upward." According to the particular style of African dance, movements can be small and subtle or large and expansive, and can include daring leaps. Most dances leave plenty of room for improvisation to the beat of the music or in dialogue with song. At their core, most African dances are about shared experience in a group. Usually a group of dancers follows the calls of one leading dancer, who calls out the next move. A group leader calling out the next step or pattern is something we also know, for example, from rueda Cubana, and from American square dance.

Do you like African music, with its emphasis on rhythm? Can you imagine being part of a group whose dancing surrenders to such rhythms and consists of expressive moves performed over a long period of time and with much stamina? Then, African dances can offer you wonderful experiences.

BELLY DANCING

Belly dances, Arab dances, the "Dance of the Seven Veils"—the style has many names and is shrouded in mystery. The belly dancing we know today was most likely created in early twentieth-century Egypt. The Middle East used to be home to many different folk dances that resembled today's belly dance but were danced mainly by men wearing harem pants. Today's belly dance is generally a solo dance performed by female dancers in specific costumes to traditional—or, sometimes, contemporary—Middle Eastern music.

The term "belly dance" is actually misleading because it reduces the complexity of Middle Eastern dancing to one body part. As in any other dance, dancers move their arms, legs, hands, feet, shoulders, and head. Belly dances are partly choreographed but leave room for improvisation. Their main movements are "shimmies"—small, quick "trembling" movements that allow the accessories of the belly-dance costume to move and, in the case of metal embellishments, to be heard. Body movements, especially of the pelvis and arms, are rhythmic and fluid. In dance class the focus is particularly on coordinating one's upper-body movements. Like all dancing, belly dancing strengthens a group's bonds. Because belly dance is often only danced by women and girls, a true sisterhood can form.

Do you like Middle Eastern music? Can you see yourself learning a solo dance, within a group, that demands intense coordination of different body parts? And would you perhaps enjoy dancing in clothes from Middle Eastern countries? If so, a belly-dancing group may be worth trying. Check it out. There are numerous opportunities for classes.

SQUARE DANCE

The origin of square dancing has not been determined conclusively, but today, square dance is widely thought of as "the happy dance with a cowboy hat." It emerged in the United States and, since the 1950s, has conquered the world.

Square dance is, of course, danced in a square formation. Four couples face each other and dance step sequences that are cued by "calls," or shouted out by a "caller." The same terms are used everywhere; once you've learned them, you can go square dancing anywhere in the world. You will work up quite a sweat, with up to 150 beats per minute, and there's always a great deal of laughter (coordination can be tricky!). Arms are usually kept close to the sides. The dance requires much from the leg muscles, and it definitely improves fitness. Generally, dancers maintain eye contact with their peers, and square dancers report that during a dance they can't help but smile. Serious addiction cannot be ruled out!

Square-dance communities in larger cities organize not only classes but also special events— good opportunities to practice, to meet fellow square dancers, to perhaps acquire a new accessory, and, above all, to dance, dance, and dance.

Do you think you could fall for country music and fast, sweat-inducing dance in a group? If you're quick and flexible enough to cope with new step combinations, you may want to try out square dancing. Most clubs offer classes for beginners.

SWING DANCE AND CO.

Swing dance is a collective term for a whole family of dances that are predominantly danced to swing music. Most prominently, they include the Charleston, Lindy Hop, jitterbug, shag, Balboa, and the blues, but also the jive, rock 'n' roll, boogie-woogie, and step dance. Even break dance and hip-hop have their roots in swing dance. Swing dance emerged in the 1920s in the ballrooms of New York and had its heyday during jazz's big-band swing era in the 1930s and '40s. Most characteristic of swing dance is its extraordinary energy, its openness, and its passion for improvisation that is so typical of jazz in general. Within the basic meter, dancers can draw freely on an almost infinite range of moves and even integrate elements from other dance traditions. Even the gender roles (leader-follower), so fixed in many other dance traditions, are not set in concrete in swing dance, and leader and follower can switch roles as they wish. The basic steps and figures in swing dancing are relatively easy to learn, and a one-day crash course is enough to get people started.

Do you like to explore new things, to improvise, and to be spontaneous? Are you fond of music from the 1920s and '30s? Do you enjoy dancing as a couple but not having your roles prescribed in detail? If you've answered these questions with "yes," you really should try out swing dancing.

CONTACT IMPROVISATION

In the early 1970s, a group of dancers developed an entirely new dance: contact improvisation, a style whose goal is to explore the maximum diversity of coordinated movements that two or more dancers can perform with each other. In addition, it emphasizes a mindful relationship with one's own body.

Typical exercises in contact improvisation are rolling, letting oneself fall, supporting another dancer's body, and letting one's weight be lifted. The movements serve the goal of bringing physical awareness to oneself and each other without language and instead letting dance become the means of communication. Aesthetic considerations play a subordinate role. The music can be quite

Dance Test: Which Style Fits Me?

varied, and sometimes movements are performed without music altogether to enable dancers to pay maximum attention to their own bodies. Contact improvisation is a group dance where everyone decides for themselves how much body contact and interaction they are willing to have with their fellow dancers.

Do you want to share experiences with others? Are you comfortable with physical closeness? Do you enjoy dancing freely without prescribed step sequences? Do you treasure mindfulness in relationships with others but also with yourself? If these questions describe you, contact improvisation might be an exciting experience. Open yourself up to it.

CLUB DANCE–FROM DISCO TO SHUFFLE

The world's first-ever disco, La Discotheque, opened in Paris during the Second World War. Because band performances were difficult to organize under the circumstances, people simply played prerecorded music to dance to. The practice spread rapidly during the 1940s, and DJs (disc jockeys) began to get people dancing in this format. Soon the first clubs that played records for people to dance to opened in London and New York City. In the 1960s and '70s, the idea of the "discotheque" spread like wildfire. Since then, people all over the world have been dancing to just about any kind of music. Mostly, the top hits of various years get played, although many hits are now produced right on the dance floors of clubs and discos; since the mid-1960s, DJs have been moving beyond simply putting on records to creating their own beats and sounds. Numerous dance moves have become famous, such as

Michael Jackson's moonwalk or, later, the shuffle dance. Since the 1990s and 2000s, house music and techno have dominated club dancing.

Do you want to celebrate life? Have fun with others and dance freely without rules? Or simply take a break, watch, chat, and drink? Then get yourself to a club!

PERSIAN DANCES

Persian dances include various folk dances, battle-stick dances, and dances for different celebrations from the many cultures and regions of modern-day Iran. There are also the dances that were practiced within the Persian courts. These dances have a complex repertoire of movements. This style is perfect for dancers who like to improvise. In contrast to belly dancing, the focus in Persian dances is on fluid arm and hand movements that often have a connection to natural phenomena or are inspired by the body language and gestures of the culture.
Persian dances are mainly about expressing the joy of living. As dancing is currently not encouraged in Iran, members of the country's diaspora find returning to these dances to be a way of connecting with Persian culture—which, after all, is one of humanity's most ancient.

You can dance Persian dances equally well to Persian music and to modern disco music. Do you have an eye for aesthetic beauty and an ear for Middle Eastern rhythms? The rhythm is different from what the Western ear is used to. Western dances are mostly to four-four or three-four rhythm, while Persian dances are danced to six-eight! Do you like the idea of learning a dance you can dance solo, with a partner, or in a group, and that is compatible with disco? Do you enjoy moving with lightness and fluidity? Then you should try out a Persian dance. Classes are offered for all age groups, and also online.

ONLINE DANCING

Are you too shy to learn to dance in public? Do you live too far away from the closest dance school to take classes regularly? Do you have responsibilities that leave you little time for yourself? Do you live in a country where dancing is forbidden? Is the world in the midst of a pandemic?

If your answer to any of the above is "yes," we have the solution for you: online dancing.

The opportunities for dance classes go beyond borders, time zones, and other limits. A little bit of online research (google "[your favorite dance style] online classes") will uncover opportunities around the world, for small groups or large, and for dancing solo, as a couple, or in your own "bubble."

Shahrzad Khorsandi and Raquel Greenberg, two experienced online-dance teachers, offer some advice for your first online dance experience. Rule #1: Be patient with yourself and with your technology. Tune in to the video call well in advance to be ready and to deal with possible technical difficulties before class starts. Rule #2: Keep a positive mindset. Online dance won't be the same as dancing in a studio. But no matter *where* you're dancing, you're doing something for your fitness, your neural connections, and for your well-being. Rule #3: Don't let anything stop you. If the music or the internet connection fails for a moment or two, *dance over it*. This will train your mental flexibility in general, and your ability to deal with frustrations. And, Rule #4: Breathe! Dance students often hold their breath. In your living room there is no teacher to remind you of breathing. Breathing gives your body and brain oxygen, and that's key for learning and for feeling good. Breathing calmly in and out also calms us down when we're stressed. So if something fails (your ability to comprehend a step, technology, the music, the internet...), just breathe deep down into the belly. And then try again.

If you're unsure of what you need for your first class, email the teacher ahead of time for advice on footwear, positioning of the video call, the best place to dance in the space you have available to you, and any other questions you may have. Some teachers may send you the musical pieces from class so you can practice offline, and some even send you a recording of the choreography after class so you can continue to practice the routine on your own until the next class.

The online-dance scene is definitely here to stay. How about tuning in to a 45-minute salsa class in between meetings? Or, to a sweaty ballet barre for beginners before dinner? Why not give it a try? We're all getting better at it with every class.

SWAYING TO MUSIC (DRINK IN HAND)

We have reached the end of this book. Maybe you're thinking: "I've looked at all the dance styles. Interesting... really interesting. But I don't feel like taking classes or going anywhere to learn to dance. No matter how healthy it's supposed to be, wild horses won't drag me there!" Never fear, we have just the right dance for you!

Remember your last night out at a bar. Or the last wedding you attended. Or even that Christmas party at your friend's

Dance Test: Which Style Fits Me?

place, when everyone pushed back the rug and turned up the music to dance. You had a lot of fun that evening (and imbibed a bit too). And then, maybe to the sounds of "Despacito" or "Uptown Funk," it happened. You linked arms with the person sitting next to you and you began to move. In time, from side to side, to the left, to the right. You did it: you swayed! Strictly speaking, you danced! And once you realize this—and realize that you enjoyed yourself—you might even pluck up the courage to make it onto the dance floor. Who knows?

We close our notebooks. The sun has set. We look at each other, satisfied and happy. Satisfied because we are able to combine our two passions, neuroscience and dancing, so beautifully. And happy because each of us has found "our" dance, which has enriched our lives and given us and continues to give us so many moments of flow.

ACKNOWLEDGMENTS

OUR FOREMOST AND big joint thank-you goes to Julia Vorrath at our publishing house, Rowohlt Verlag, to Andy Hartard at HERBERT Management, and to Daniel Mursa at Agentur Petra Eggers. Many, many thanks for your infinite trust in us.

We also thank our colleagues at the world's best conference, on the Greek island of Aegina, where we first met: especially Professors Manos Tsakiris, Ophelia Deroy, Barry Smith, and Katarina Fotopoulou, but also all the other participants.

Many people have shared my, Julia's, journey. I especially thank my parents, Inge-Lise and Gerhard. My share of this book's dedication is for the two of you. I could not have done it without you. Love ya...

Another thank-you goes to my friend, almost sister, Nadine, who has supported me emotionally during this book-journey, and to Mattia.

Needless to say, I am grateful to Dong-Seon for the wonderful opportunity to write this book with him. I think we made the impossible possible! Thank you for trusting me.

Acknowledgments

Many thanks also to my colleagues, who stand by me like friends, laugh at my outdated jokes, and encourage me when I have doubts. They include, among others, Professor Antoni Gomila, Professor Shelley Channon, Professor Vincent Walsh, Dr. H.E., Professor Joseph Devlin, and Knud Engsnap, inimitable teacher at the Duburg Skolen in Flensburg. I also thank my dear friends Isa Soriano, Miguel Azorin, and Melanie Ring, as well as my exuberant colleagues at the Applied Consumer Neuroscience Company in London, our "Persian/Iranian Dance-Science" team, and many others.

I am grateful to my many dance friends, especially Robert Hovey, Valerie Strobel, Ginny Möller, and the London tango gang. Also, to the teams of Negracha Milonga, Etnia, Corrientes, Yumba Milonga, La Divina with Raquel Greenberg, and Bellezza Milonga, and to the many, many other dancers who sweetened my nights with dance—you know who you are.

I also thank all who have made life difficult for me. Every day you make me stronger, even if that is not your intention. You, too, know who you are.

Some of this book was written in cafés around Europe. Many thanks to the wonderful staff at Ca'n Joan de S'Aigo, Amadip Esment, and S'Estació (Palma, Spain), Belle Epoque and Blend (London, UK), Café Guldægget (Esbjerg, Denmark), Das Schiffercafé and Peaberries (Kiel, Germany), and the Pasticceria via Correggio (Milano, Italy). I am also grateful to the hot yoga teams in Palma, Milano-Navigli, Kiel, and London for a "new back."

For me, Dong-Seon, this book is above all an attempt to persuade my wife, Eugene, to dance with me. To feel the closeness of the person we love and to forget everything around us while we dance together is a wonderful experience that I would love to share with her. One of my dreams is for us to grow old as a dancing couple. This is why I dedicate this book to you, Eugene.

Writing a book is always a big undertaking that requires much strength and energy. I couldn't have done it had my family not been so understanding and willing to let me have so much time we otherwise could have spent together. Hence my first thank-you goes to my family—my wife, Eugene, and my son, Theo.

A special thank-you goes to my co-author, Julia, without whom this book would not have been possible. That we were able to complete it is above all a credit to her knowledge, her passion, her ability to motivate others, and her dedication.

Because this is a book about dance, I would like to thank everyone who has led me to dancing: Jeannette Neustadt, with whom I had the opportunity to direct beautiful dance performances and dance-theater shows and who was the first to introduce me to the dance theater by Pina Bausch. To Tessa Theodorakopoulos and Konstantin Tsakalidis, for whose theater and dance shows I had the privilege of composing the music. To my fellow students at Konstanz University, Germany, who gave me my very first crash course in ballroom dances so I could attend the university ball. To my Mexican, Colombian, and Puerto Rican roommates at Rutgers University, with whom I danced my first salsa. And to my colleagues and fellow students at the Max Planck

Institute for Biological Cybernetics, with whom I also practiced salsa dancing (and where I did research on salsa). Also, I'd like to thank the unforgettable Ringopop family in Korea, where I learned to dance swing, and Shin, who brought me to swing dancing.

Adamas, BB Mingyu Kim, Soochan Lee, Hyunjeong Choi, Rico Lim and Chloe Hong, Thomas Blacharz and Alice Mei were the first teachers who introduced me to dance. Many thanks to the colleagues of Swingzeit Tübingen with whom I taught swing in the university sports program and built a swing scene in Tübingen. I'd like to also thank Swingkultur and Swingstep Stuttgart who helped us a lot in Tübingen. And thank you, all you Lindy Hoppers in Berlin, Munich, Hamburg, Cologne, and Bonn for being so welcoming each time I came for a visit. There are many more dancers in the swing scenes of the world that I would like to thank for giving me wonderful dance moments.

Many thanks also to the science slammers who were always so supportive of me: Dennis Fink and Johannes von Borstel.

Last but not least, thank you to the cafés where I spent so much time writing: Ca'n Joan de S'Aigo (Palma, Spain), Café Moon, and Toz Study Café (Korea). Thank you!

SOURCES

CHAPTER 1

Arnal, L.H., et. al. "Human Screams Occupy a Privileged Niche in the Communication Soundscape." *Current Biology* 25, no. 15 (August 3, 2015): 2051–56.

Ayotte, J., et al. "Congenital Amusia: A Group Study of Adults Afflicted With a Music-Specific Disorder." *Brain* 125, pt. 2 (2002): 238–51.

Buss, A.H., et al. "The Development of Embarrassment." *Journal of Psychology* 103 (1979): 227–30.

Buzsáki, G. *Rhythms of the Brain.* New York: Oxford University Press, 2009.

Buzsáki, G., and A. Draguhn. "Neuronal Oscillations in Cortical Networks." *Science* 304, no. 5679 (2004): 1926–29.

Christensen, et al. "Affective Responses to Dance." *Acta Psychologica* 168 (2016): 91–105.

Cupchik, G.C. "The Legacy of Daniel E. Berlyne." *Empirical Studies of the Arts* 6, no. 2 (1988): 171–86.

Ekardt, P. "Certain Wonderful Gestures: Warburg, Lessing and the Transitory in Images." *Culture, Theory and Critique* 57, no. 2 (2016): 166–75.

Ekman, P. "An Argument for Basic Emotions." *Cognition and Emotion* 6, no. 3–4 (1992): 169–200.

Ekman, P. "Basic Emotions." In *Handbook of Cognition and Emotion*, edited by T. Dalgleish and M.J. Power, 45–60. New York: John Wiley & Sons, 1999.

Ernst, M.O. "A Bayesian View on Multimodal Cue Integration." In *Advances in Visual Cognition: Human Body Perception From the Inside Out*, edited by G. Knoblich, I.M. Thornton, M. Grosjean, and M. Shiffrar, 105–31. New York: Oxford University Press, 2006.

Fleischhauer, W. *Drei Minuten mit der Wirklichkeit*. Munich: Droemer-Knaur, 2002.

Gardner, H. "Review: Aesthetics and Psychobiology by D. E. Berlyne." *Curriculum Theory Network* 4, no. 2/3 (1974): 205–11.

Gilovich, T., and K. Savitsky. "The Spotlight Effect and the Illusion of Transparency: Egocentric Assessments of How We Are Seen by Others." *Current Directions in Psychological Science* 8, no. 6 (1999): 165–68.

Holmes, E.A., and A. Mathews. "Mental Imagery and Emotion: A Special Relationship?" *Emotion* 5, no. 4 (2005): 489–97.

Huron, D. "Is Music an Evolutionary Adaptation?" In *The Cognitive Neuroscience of Music*, edited by I. Peretz and R. Zatorre, 43–61. Oxford: Oxford University Press, 2006.

Huron, D. "Sweet Anticipation: Music and the Psychology of Expectation." *Music Perception* 24, no. 5 (2007): 511–14.

Körding, K.P., et al. "Causal Inference in Multisensory Perception." PLOS ONE 2, no. 9 (2007).

Kreutz, G., and M. Lotze. "Neuroscience of Music and Emotion." In *Neurosciences in Music Pedagogy*, edited by W. Gruhn and F. Rauscher, 143–67. New York: Nova Biomedical Books, 2007.

Kreutz, G., et al. "Using Music to Induce Emotions: Influences of Musical Preference and Absorption." *Psychology of Music* 36, no. 1 (2008): 101–26.

MacDonald, R., et al. *Music, Health, and Wellbeing*. Oxford: Oxford University Press, 2013.

Monaghan, T. "Why Study the Lindy Hop?" *Dance Research Journal* 33, no. 2 (2001): 124–27.

Neal, D.T., and T.L. Chartrand. "Embodied Emotion Perception: Amplifying and Dampening Facial Feedback Modulates Emotion Perception Accuracy." *Social Psychological and Personality Science* 2, no. 6 (2011): 673–78.

Niedenthal, P.M. "Embodying Emotion." *Science* 316, no. 5827 (2007): 1002–1005.

Parise, C.V., et al. "When Correlation Implies Causation in Multi-Sensory Integration." *Current Biology* 22, no. 1 (2012): 46–49.

Peretz, I., and D.T. Vuvan. "Prevalence of Congenital Amusia." *European Journal of Human Genetics* 25, no. 5 (2017): 625–30.

Phillips-Silver, J., et al. "Born to Dance but Beat Deaf: A New Form of Congenital Amusia." *Neuropsychologia* 49, no. 5 (2011): 961–69.

Prinz, J., and P. James. "Emotions Embodied." In *Blackwell Philosophy Anthologies. Mind and Cognition: An Anthology*, edited by W.G. Lycan and J.J. Prinz, 839–49. New Jersey: Blackwell Publishing, 2008.

Quiroga Murcia, C., et al. "Shall We Dance? An Exploration of the Perceived Benefits of Dancing on Well-Being." *Arts & Health* 2, no. 2 (2010): 149–63.

Savitsky, K., et al. "Do Others Judge Us as Harshly as We Think? Over-Estimating the Impact of Our Failures, Shortcomings, and Mishaps." *Journal of Personality and Social Psychology* 81, no. 1 (2001): 44–56.

Savitsky, K., and T. Gilovich. "The Illusion of Transparency and the Alleviation of Speech Anxiety. *Journal of Experimental Social Psychology* 39, no. 6 (2003): 618–25.

Stewart, L. "Fractionating the Musical Mind: Insights From Congenital Amusia." *Current Opinion in Neurobiology* 18, no. 2 (2008): 127–30.

Warburg, A., and T. Bodart. *Le rituel du serpent: récit d'un voyage en pays pueblo. La littérature artistique.* Paris: Macula, 2003.

Warburg, A., and W.F. Mainland. "A Lecture on Serpent Ritual." *Journal of the Warburg Institute* 2, no. 4 (1939).

Winkler, I., et al. "Newborn Infants Detect the Beat in Music." *Proceedings of the National Academy of Sciences* 106, no. 7 (2009): 2468–71.

Witek, M.A.G. "Filling In: Syncopation, Pleasure and Distributed Embodiment in Groove." *Music Analysis* 36, no. 1 (2017): 138–60.

Witek, M.A.G., et al. "Syncopation, Body-Movement and Pleasure in Groove Music." *PLOS ONE* 10, no. 9 (2014).

Zentner, M., and T. Eerola. "Rhythmic Engagement With Music in Infancy." *Proceedings of the National Academy of Sciences* 107, no. 13 (2010): 5768–73.

CHAPTER 2

Allport, G.W. *Personality: A Psychological Interpretation.* New York: Henry Holt, 1937.

Allport, G.W. "Traits Revisited." *American Psychologist* 21, no. 1 (1966): 1–10.

Bensafi, M., et al. "Sex-Steroid Derived Compounds Induce Sex-Specific Effects on Autonomic Nervous System Function in Humans." *Behavioral Neuroscience* 117, no. 6 (2003): 1125–34.

Bernhardt, B.C., and T. Singer. "The Neural Basis of Empathy." *Annual Review of Neuroscience* 35 (2012): 1–23.

Boone, R.T., and J.G. Cunningham. "Children's Expression of Emotional Meaning in Music Through Expressive Body Movement." *Journal of Nonverbal Behavior* 25 (2001): 21–41.

Boulenger, V., et al. "Grasping Ideas With the Motor System: Semantic Somatotopy in Idiom Comprehension." *Cerebral Cortex* 19, no. 8 (2009): 1905–14.

Brook, P. "The Deadly Theatre." In *The Empty Space*, 7–48. New York: Touchstone, 1968.

Brown, S., M.J. Martinez, and L.M. Parsons. "The Neural Basis of Human Dance." *Cerebral Cortex* 16 (2006): 1157–67.

Brown, S., and L.M. Parsons. "The Neuroscience of Dance." *Scientific American* 299, no. 1 (2008): 78–83.

Brown, W.M., et al. "Dance Reveals Symmetry Especially in Young Men." *Nature* 438 (2005): 1148–50.

Brück, C., et al. "Emotional Voices in Context: A Neurobiological Model of Multimodal Affective Information Processing." *Physics of Life Reviews* 8, no. 4 (2011): 383–403.

Calvo-Merino, B., et al. "Action Observation and Acquired Motor Skills: An fMRI Study With Expert Dancers." *Cerebral Cortex* 18, no. 8 (2005): 1243–49.

Chang, D.-S., et al. "Actions Revealing Cooperation: Predicting Cooperativeness in Social Dilemmas From the Observation of Everyday Actions." *Cognitive Processing* 15, no. 1 (2014).

Chauvigné, L.A.S., et al. "Taking Two to Tango: fMRI Analysis of Improvised Joint Action With Physical Contact." *PLOS ONE* 13, no. 1 (2018).

Christensen, J.F., et al. "Dance Expertise Modulates Behavioral and Psychophysiological Responses to Affective Body Movement." *Journal of Experimental Psychology: Human Perception and Performance* 42, no. 8 (2016): 1139–47.

Christensen, J.F., et al. "Not All About Sex? Neural and Biobehavioural Functions of Human Dance." *Proceedings of the New York Academy of Sciences* 1400, no. 1 (2017): 8–32.

Coates, J.M., et al. "Second-to-Fourth Digit Ratio Predicts Success Among High-Frequency Financial Traders." *Proceedings of the National Academy of Sciences* 106, no. 2 (2009): 623–28.

Croyden, M. *Conversations With Peter Brook, 1970–2000.* New York: Theatre Communications Group, 2003.

Darwin, C. *The Expression of Emotion in Man and Animals*. Cambridge, MA: Cambridge University Press, 2013.

de Vignemont, F., and T. Singer. "The Empathic Brain: How, When and Why?" *Trends in Cognitive Sciences* 10, no. 10 (2006): 435–41.

di Pellegrino, G., et al. "Understanding Motor Events: A Neurophysiological Study." *Experimental Brain Research* 91 (1992): 176–80.

Ebberfeld, I. *Botenstoffe der Liebe: Über das innige Verhältnis von Geruch und Sexualität*. Munster: LIT Verlag, 2005.

Fedorov, L., et al. "Adaptation Aftereffects Reveal Representations for Encoding of Contingent Social Actions." *Proceedings of the National Academy of Sciences* 115, no. 29 (2018): 7515–20.

Fink, B., et al. "A Preliminary Investigation of the Associations Between Digit Ratio and Women's Perception of Men's Dance." *Personality and Individual Differences* 42, no. 2 (2007): 381–90.

Fink, B., et al. "Second to Fourth Digit Ratio and the 'Big Five' Personality Factors." *Personality and Individual Differences* 37, no. 3 (2004): 495–503.

Fink, B., et al. "Second to Fourth Digit Ratio and Sensation Seeking." *Personality and Individual Differences* 41, no. 7 (2006): 1253–62.

Fink, B., et al. "Women's Body Movements Are a Potential Cue to Ovulation." *Personality and Individual Differences* 53, no. 6 (2012): 759–63.

Gallese, V., et al. "Action Recognition in the Premotor Cortex." *Brain* (1996, part 2): 593–609.

Glenberg, A.M. "Language and Action: Creating Sensible Combinations of Ideas." In *The Oxford Handbook of Psycholinguistics*, edited by M.G. Gaskell. Oxford University Press, 2012.

Glenberg, A.M., and V. Gallese. "Action-Based Language: A Theory of Language Acquisition, Comprehension, and Production." *Cortex* 48, no. 7 (2012): 905–22.

Glenberg, A.M., and M.P. Kaschak. "Grounding Language in Action." *Psychonomic Bulletin & Review* 9 (2002): 558–65.

Hasson, U. "This Is Your Brain on Communication." Filmed February 2016 in Vancouver, Canada. TED video.

Hasson, U., et al. "Brain-to-Brain Coupling: A Mechanism for Creating and Sharing a Social World." *Trends in Cognitive Sciences* 16, no. 2 (2012): 114–21.

Hauk, O., et al. "Somatotopic Representations of Action Words in Muman Motor and Premotor Cortex." *Neuron* 41, no. 2 (2004): 301–7.

Havlicek, J., and S.C. Roberts. "MHC-Correlated Mate Choice in Humans: A Review." *Psychoneuroendocrinology* 34, no. 4 (2009): 497–512.

Hejmadi, A., et al. "Exploring Hindu Indian Emotion Expressions: Evidence for Accurate Recognition by Americans and Indians." *Psychological Science* 11, no. 3 (2000): 183–87.

Hess, E.H., and J.M. Polt. "Pupil Size as Related to Interest Value of Visual Stimuli." *Science* 132, no. 3423 (1960): 349–50.

Hugill, N., et al. "Women's Perception of Men's Sensation Seeking Propensity From Their Dance Movements." *Personality and Individual Differences* 51, no. 4 (2011): 483–87.

Kellerman, J., et al. "Looking and Loving: The Effects of Mutual Gaze on Feelings of Romantic Love." *Journal of Research in Personality* 23, no. 2 (1989): 145–61.

Kempel, P., et al. "Second-to-Fourth Digit Length, Testosterone and Spatial Ability." *Intelligence* 33, no. 3 (2005): 215–30.

Kreifelts, B., et al. "They Are Laughing at Me: Cerebral Mediation of Cognitive Biases in Social Anxiety." *PLoS ONE* 9, no. 6 (2014).

Laird, J.D. *Feelings: The Perception of Self.* New York: Oxford University Press, 2007.

Laird, J.D., and K. Lacasse. "Bodily Influences on Emotional Feelings: Accumulating Evidence and Extensions of William James's Theory of Emotion." *Emotion Review* 6, no. 1 (2014): 27–34.

Lawlor, L., and V. Moulard Leonard. "Henri Bergson." In *The Stanford Encyclopaedia of Philosophy*, edited by E.N. Zalta. Summer 2016 edition.

Lovatt, P. "Dance Confidence, Age and Gender." *Personality and Individual Differences* 50, no. 5 (2011): 668–72.

Lovatt, P. *Dance Psychology.* Norfolk, UK: Dance Presents, 2018.

MacDonald, R., et al. *Music, Health, and Wellbeing.* Oxford: Oxford University Press, 2012.

Manning, J.T., and R.P. Taylor. "Second to Fourth Digit Ratio and Male Ability in Sport: Implications for Sexual Selection in Humans." *Evolution and Human Behavior* 2, no. 1 (2001): 61–69.

Mayer, K.M., et al. "Visual and Motor Cortices Differentially Support the Translation of Foreign Language Words." *Current Biology* 25, no. 4 (2015): 530–35.

McCarty, K., et al. "Male Body Movements as Possible Cues to Physical Strength: A Biomechanical Analysis." *American Journal of Human Biology* 25, no. 3 (2013): 307–12.

Miller, G., et al. "Ovulatory Cycle Effects on Tip Earnings by Lap Dancers: Economic Evidence for Human Estrus?" *Evolution and Human Behavior* 28, no. 6 (2007): 375–81.

Millet, K., and S. Dewitte. "Second to Fourth Digit Ratio and Cooperative Behavior." *Biological Psychology* 71, no. 1 (2006): 111–15.

Neal, D.T., and T.L. Chartrand. "Embodied Emotion Perception: Amplifying and Dampening Facial Feedback Modulates Emotion Perception Accuracy." *Social Psychological and Personality Science* 2, no. 6 (2011): 673–78.

Neave, N., et al. "Male Dance Moves That Catch a Woman's Eye." *Biology Letters* 7 (2011): 221–24.

Neave, N., et al. "Second to Fourth Digit Ratio, Testosterone and Perceived Male Dominance." *Proceedings of the Royal Society B: Biological Sciences* 270 (2003): 2167–72.

Nicolescu, B., and D. Williams. "Peter Brook and Traditional Thought." *Contemporary Theatre Review* 7, no. 1 (1998): 11–23.

Noy, L., et al. "The Mirror Game as a Paradigm for Studying the Dynamics of Two People Improvising Motion Together." *Proceedings of the National Academy of Sciences* 108, no. 52 (2011): 20947–52.

Partala, T., and V. Surakka. "Pupil Size Variation as an Indication of Affective Processing." *International Journal of Human-Computer Studies* 59, no. 1–2 (2003): 185–98.

Pulvermüller, F. "How Neurons Make Meaning: Brain Mechanisms for Embodied and Abstract-Symbolic Semantics." *Trends in Cognitive Sciences* 17, no. 9 (2013): 458–70.

Pulvermüller, F. "Meaning and the Brain: The Neurosemantics of Referential, Interactive, and Combinatorial Knowledge." *Journal of Neurolinguistics* 25, no. 5 (2012): 423–59.

Pulvermüller, F. "Semantic Embodiment, Disembodiment or Misembodiment? In Search of Meaning in Modules and Neuron Circuits." *Brain and Language* 127, no. 1 (2013): 86–103.

Pulvermüller, F., and L. Fadiga. "Active Perception: Sensorimotor Circuits as a Cortical Basis for Language." *Nature Reviews Neuroscience* 11 (2010): 351–60.

Rizzolatti, G., and L. Craighero. "The Mirror-Neuron System." *Annual Review of Neuroscience* 27 (2004): 169–92.

Rizzolatti, G., et al. "Neurophysiological Mechanisms Underlying the Understanding and Imitation of Action." *Nature Review Neuroscience* 2, no. 9 (2001): 661–70.

Rizzolatti, G., and C. Sinigaglia. "The Functional Role of the Parietofrontal Mirror Circuit: Interpretations and Misinterpretations." *Nature Reviews Neuroscience* 11, no. 4 (2010): 264–74.

Santos, P.S., et al. "New Evidence That the MHC Influences Odor Perception in Humans: A Study With 58 Southern Brazilian Students." *Hormones and Behavior* 47, no. 4 (2005): 384–88.

Shebani, Z., and F. Pulvermüller. "Moving the Hands and Feet Specifically Impairs Working Memory for Arm- and Leg-Related Action Words." *Cortex* 49, no. 1 (2013): 222–31.

Shevtsova, M. "Interculturalism, Aestheticism, Orientalism: Starting From Peter Brook's Mahabharata." *Theatre Research International* 22, no. 2 (1997): 98–104.

Singer, T., and C. Lamm. "The Social Neuroscience of Empathy." *Annals of the New York Academy of Sciences* 1156 (2009): 81–96.

Singer, T., et al. "Empathic Neural Responses Are Modulated by the Perceived Fairness of Others." *Nature* 439 (2006): 466–69.

Smith, A. *The Theory of Moral Sentiments*. London, 1759.

Wedekind, C., et al. "MHC-Dependent Mate Preferences in Humans." *Proceedings of the Royal Society B: Biological Sciences* 260, no. 1359 (1995): 245–49.

Wolpert, D. "The Real Reason for Brains." Filmed July 2011 in Edinburgh, UK. TED video.

Wolpert, D.M., et al. "A Unifying Computational Framework for Motor Control and Social Interaction." *Philosophical Transactions of the Royal Society of London. Series B, Biological Sciences* 358, no. 1431 (2003): 593–602.

Wolpert, D.M., and Z. Ghahramani. "Computational Principles of Movement Neuroscience." *Nature Neuroscience* 3 (2000): 1212–17.

Wunderer, E., and K.A. Schneewind. "The Relationship Between Marital Standards, Dyadic Coping and Marital Satisfaction." *European Journal of Social Psychology* 38, no. 3 (2008): 462–76.

Wunderer, E., et al. "Ehebeziehungen: Eine Typologie auf Basis von Paarklima-Skalen." *Zeitschrift Für Familienforschung* (2001): 74–95.

CHAPTER 3

Adolphs, R. "Social Cognition and the Human Brain." *Trends in Cognitive Sciences* 3, no. 12 (1999): 469–79.

Altinok, T. *Der Volkstanz als Prozess des interkulturellen Lernens. Eine explorative Studie*. Cologne: German Sport University, 2011.

Amodio, D.M. "The Social Neuroscience of Intergroup Relations." *European Review of Social Psychology* 19 (2008): 1–54.

Cacioppo, J.T., et al. "Evolutionary Mechanisms for Loneliness." *Cognition and Emotion* 28, no. 1 (2014): 3–21.

Cacioppo, J.T., and L.C. Hawkley. "Perceived Social Isolation and Cognition." *Trends in Cognitive Sciences* 13, no. 10 (2009): 447–54.

Cacioppo, J.T., et al. "The Anatomy of Loneliness." *Current Directions in Psychological Science* 12, no. 3 (2003): 71–74.

Cacioppo, J.T., et al. (2006). "Loneliness as a Specific Risk Factor for Depressive Symptoms: Cross-Sectional and Longitudinal Analyses." *Psychology and Aging* 21, no. 1 (2006): 140–51.

Cheon, B.K., et al. "Cultural Influences on Neural Basis of Inter-Group Empathy." *NeuroImage* 57, no. 2 (2011): 642–50.

D'Ausilio, A., et al. "Leadership in Orchestra Emerges From the Causal Relationships of Movement Kinematics." PLOS ONE 7, no. 5 (2012).

Desmond, J., ed. *Meaning in Motion: New Cultural Studies of Dance*. *American Anthropologist*. Durham, NC: Duke University Press, 1997.

Ernst, J.M., and J.T. Cacioppo. "Lonely Hearts: Psychological Perspectives on Loneliness." *Applied and Preventive Psychology* 8, no. 1 (1999): 1–22.

Haidt, J., et al. "Hive Psychology, Happiness, and Public Policy." *Journal of Legal Studies* 37 (2008): S133–56.

Heinrich, L.M., and E. Gullone. "The Clinical Significance of Loneliness: A Literature Review." *Clinical Psychology Review* 26, no. 6 (2006): 695–718.

Kanai, R., et al. "Brain Structure Links Loneliness to Social Perception." *Current Biology* 22, no. 20 (2012): 1975–79.

Keyfitz, N., and W.H. McNeill. "Keeping Together in Time: Dance and Drill in Human History." *Contemporary Sociology* 25, no. 3 (1996): 408–9.

Klucharev, V., et al. "Reinforcement Learning Signal Predicts Social Conformity." *Neuron* 61, no. 1 (2009): 140–51.

Launay, J., et al. "Synchrony as an Adaptive Mechanism for Large-Scale Human Social Bonding." *Ethology* 122, no. 10 (2016): 779–89.

Marsh, K.L., et al. "Social Connection Through Joint Action and Interpersonal Coordination." *Topics in Cognitive Science* 1, no. 2 (2009): 320–39.

Masi, C.M., et al. "A Meta-Analysis of Interventions to Reduce Loneliness." *Personality and Social Psychology Review* 15, no. 3 (2011): 219–66.

Mathur, V., et al. "Neural Basis of Extraordinary Empathy and Altruistic Motivation." *NeuroImage* 51, no. 4 (2010): 1468–75.

Meltzoff, A.N., and J. Decety. "What Imitation Tells Us About Social Cognition: A Rapprochement Between Developmental Psychology and Cognitive Neuroscience." *Philosophical Transactions of the Royal Society of London. Series B, Biological Sciences* 358, no. 1431 (2003): 491–500.

Orgs, G., et al. "Expertise in Dance Modulates Alpha/Beta Event-Related Desynchronization During Action Observation." *European Journal of Neuroscience* 27, no. 12 (2008): 3380–84.

Orgs, G., and P. Haggard. "Temporal Binding During Apparent Movement of the Human Body." *Visual Cognition* 19, no. 7 (2011): 833–45.

Orgs, G., et al. "Learning to Like It: Aesthetic Perception of Bodies, Movements and Choreographic Structure." *Consciousness and Cognition* 22, no. 2 (2013): 603–12.

Orgs, G., et al. "Time Perception During Apparent Biological Motion Reflects Subjective Speed of Movement, Not Objective Rate of Visual Stimulation." *Experimental Brain Research* 227, no. 2 (2013): 223–29.

Rand, D.G., and M. Nowak. "Human Cooperation." *Trends in Cognitive Sciences* 17, no. 8 (2013): 413–25.

Riketta, M. "Cognitive Differentiation Between Self, Ingroup, and Outgroup: The Roles of Identification and Perceived Intergroup Conflict." *European Journal of Social Psychology* 35, no. 1 (2005): 97–106.

Rilling, J.K., et al. "The Neural Correlates of the Affective Response to Unreciprocated Cooperation." *Neuropsychologia* 46, no. 5 (2008): 1256–66.

Steptoe, A., et al. "Social Isolation, Loneliness, and All-Cause Mortality in Older Men and Women." *Proceedings of the National Academy of Sciences of the United States of America* 110, no. 15 (2013): 5797–5801.

Stevens, J.R., and M.D Hauser. "Why Be Nice? Psychological Constraints on the Evolution of Cooperation." *Trends in Cognitive Sciences* 8, no. 2 (2004): 60–65.

Stupacher, J., et al. "Music Strengthens Prosocial Effects of Interpersonal Synchronization—If You Move in Time With the Beat." *Journal of Experimental Social Psychology* 72 (2017): 39–44.

Tarr, B., et al. "Synchrony and Exertion During Dance Independently Raise Pain Threshold and Encourage Social Bonding." *Biology Letters* 11, no. 10 (2015).

Tarr, B., et al. "Music and Social Bonding: 'Self-Other' Merging and Neurohormonal Mechanisms." *Frontiers in Psychology* 5 (2014): 1096.

Tarr, B., et al. "Silent Disco: Dancing in Synchrony Leads to Elevated Pain Thresholds and Social Closeness." *Evolution and Human Behavior* 37, no. 5 (2016): 343–49.

Tomasello, M., and A. Vaish. "Origins of Human Cooperation and Morality." *Annual Review of Psychology* 64, no. 1 (2012): 231–55.

Valdesolo, P., and D. Desteno. "Synchrony and the Social Tuning of Compassion." *Emotion* 11, no. 2 (2011): 262–66.

Vicary, S., et al. "Joint Action Aesthetics." *PLOS ONE* 12, no. 7 (2017).

von Zimmermann, J., et al. "The Choreography of Group Affiliation." *Topics in Cognitive Science* 10, no. 1 (2018): 80–94.

Walton, G.M., et al. "Mere Belonging: The Power of Social Connections." *Journal of Personality and Social Psychology* 102, no. 3 (2012): 513–32.

Wiltermuth, S.S., and C. Heath. "Synchrony and Cooperation." *Psychological Science* 20, no. 1 (2009): 1–5.

CHAPTER 4

Baker, F.A., and R.A.R MacDonald. "Flow, Identity, Achievement, Satisfaction and Ownership During Therapeutic Songwriting Experiences With University Students and Retirees." *Musicae Scientiae* 17, no. 2 (2013): 129–44.

Barrett, K.C., et al. "Art and Science: How Musical Training Shapes the Brain." *Frontiers in Psychology* 4 (2013): 713.

Barrett, L.F., et al. "Interoceptive Sensitivity and Self-Reports of Emotional Experience." *Journal of Personality and Social Psychology* 87, no. 5 (2004): 684–97.

Berridge, K.C., and M.L. Kringelbach. "Pleasure Systems in the Brain." *Neuron* 86, no. 3 (2015): 646–64.

Bläsing, B., et al. "Neurocognitive Control in Dance Perception and Performance." *Acta Psychologica* 139, no. 2 (2012): 300–308.

Bradt, J., et al. "Music for Stress and Anxiety Reduction in Coronary Heart Disease Patients." *Cochrane Database of Systematic Reviews* 2013, Issue 12.

Bufalari, I., and S. Ionta, S. "The Social and Personality Neuroscience of Empathy for Pain and Touch." *Frontiers in Human Neuroscience* 7 (2013): 393.

Burzynska et al. "The Dancing Brain: Structural and Functional Signatures of Expert Dance Training." *Frontiers in Human Neuroscience* 11 (2017): 566.

Cacioppo, S., and J.T. Cacioppo. "Decoding the Invisible Forces of Social Connections." *Frontiers in Integrative Neuroscience* 6 (2012): 51.

Cameron, O.G. "Interoception: The Inside Story—a Model for Psychosomatic Processes." *Psychosomatic Medicine* 63, no. 5 (2001): 697–710.

Cassidy, G., and R.A.R Macdonald. "The Effect of Background Music and Background Noise on the Task Performance of Introverts and Extraverts." *Psychology of Music* 35, no. 3 (2007): 517–37.

Castrillon, T. et al. "The Effects of a Standardized Belly Dance Program on Perceived Pain, Disability, and Function in Women With Chronic Low Back Pain." *Journal of Back and Musculoskeletal Rehabilitation* 30, no. 3 (2017): 1–20.

Cepeda, C.C.P., et al. "Effect of an Eight-Week Ballroom Dancing Program on Muscle Architecture in Older Adult Females." *Journal of Aging and Physical Activity* 23, no. 4 (2015).

Cherches, I.M. "Clinical Neuroanatomy." In *Neurology Secrets: Sixth Edition*, edited by J. Kass and E. Mizrahi, 11–41. Philadelphia: Elsevier, 2017.

Christensen, J.F. "Pleasure Junkies All Around! Why It Matters and Why 'The Arts' Might Be the Answer." *Proceedings of the Royal Society B: Biological Sciences* 284, no. 1854 (2017).

Christensen, J.F., et al. "I Can Feel My Heartbeat: Dancers Have Increased Interoceptive Awareness." *Psychophysiology* 55, no. 4 (2018).

Clayton, N. "Dancing to Darwin." *Current Biology* 19, no. 17 (2009): R725.

Clift, S. "Creative Arts as a Public Health Resource: Moving From Practice-Based Research to Evidence-Based Practice." *Perspectives in Public Health* 132, no. 3 (2012): 120–27.

Corbetta, D., and W. Snapp-Childs. "Seeing and Touching: The Role of Sensory-Motor Experience on the Development of Infant Reaching." *Infant Behavior and Development* 32, no. 1 (2009): 44–58.

Craig, A.D. "How Do You Feel? Interoception: The Sense of the Physiological Condition of the Body." *Nature Reviews. Neuroscience* 3, no. 8 (2002): 655–66.

Craig, A. "Interoception: The Sense of the Physiological Condition of the Body." *Current Opinion in Neurobiology* 13, no. 4 (2003): 500–505.

Critchley, H.D. "Psychophysiology of Neural, Cognitive and Affective Integration: fMRI and Autonomic Indicants." *International Journal of Psychophysiology: Official Journal of the International Organization of Psychophysiology* 73, no. 2 (2009): 88–94.

Critchley, H.D., and Y. Nagai. "How Emotions Are Shaped by Bodily States." *Emotion Review* 4, no. 2 (2012): 163–68.

Critchley, H.D., et al. "Neural Systems Supporting Interoceptive Awareness." *Nature Neuroscience* 7 (2004): 189–94.

Csikszentmihalyi, M. *Flow: The Classic Work on How to Achieve Happiness.* London: Rider, 2002.

Csikszentmihalyi, M. *Flow: The Psychology of Happiness.* New York: Harper & Row, 1990.

Csikszentmihalyi, M. *Flow: The Psychology of Optimal Experience.* New York: HarperCollins, 1991.

Dunbar, R.I.M. "The Social Role of Touch in Humans and Primates: Behavioral Function and Neurobiological Mechanisms." *Neuroscience and Biobehavioral Reviews* 34, no. 2 (2010): 260–68.

Dunn, B.D., et al. "Listening to Your Heart: How Interoception Shapes Emotion, Experience, and Intuitive Decision Making." *Psychological Science* 21, no. 12 (2010): 1835–44.

Ernst, J., et al. "Interoceptive Awareness Enhances Neural Activity During Empathy." *Human Brain Mapping* 34, no. 7 (2013): 1615–24.

Esch, T., and E. von Hirschhausen. *Die Neurobiologie des Glücks. Wie die Positive Psychologie die Medizin verändert.* Leipzig: Thieme Verlag, 2014.

Ferri, F., et al. "Closing the Gap Between the Inside and the Outside: Interoceptive Sensitivity and Social Distances." *PLOS ONE* 8, no. 10 (2013).

Ferri, F., et al. "When Action Meets Emotions: How Facial Displays of Emotion Influence Goal-Related Behavior." *PLOS ONE* 5, no. 10 (2010).

Fitch, W.T. "The Evolution of Music in Comparative Perspective." *Annals of the New York Academy of Sciences* 1060 (2005): 29–49.

Fukui, H. "Efficacy of Music Therapy in Treatment for the Patients With Alzheimer's Disease." *International Journal of Alzheimer's Disease* (2012).

Fukui, H., and K. Toyoshima. "Music Facilitate the Neurogenesis, Regeneration and Repair of Neurons." *Medical Hypotheses* 71, no. 5 (2008): 765–69.

Fukushima, H., et al. "Association Between Interoception and Empathy: Evidence From Heartbeat-Evoked Brain Potential." *International Journal of Psychophysiology* 79, no. 2 (2011): 259–75.

Gallace, A., and C. Spence. "The Science of Interpersonal Touch: An Overview." *Neuroscience and Biobehavioral Reviews* 34, no. 2 (2010): 246–59.

Garlin, F.V., and K. Owen. "Setting the Tone With the Tune: A Meta-Analytic Review of the Effects of Background Music in Retail Settings." *Journal of Business Research* 59, no. 6 (2006): 755–64.

Gaser, C., and G. Schlaug. "Brain Structures Differ Between Musicians and Non-Musicians." *The Journal of Neuroscience* 23, no. 27 (2003): 9240–45.

Gerra, G., et al. "Neuroendocrine Responses of Healthy Volunteers to 'Techno-Music': Relationships With Personality Traits and Emotional State." *International Journal of Psychophysiology: Official Journal of the International Organization of Psychophysiology* 28, no. 1 (1998): 99–111.

Guéguen, N. "Courtship Compliance: The Effect of Touch on Women's Behavior." *Social Influence* 2, no. 2 (2007): 81–97.

Guéguen, N. "The Effect of a Woman's Incidental Tactile Contact on Men's Later Behavior." *Social Behavior and Personality: An International Journal* 38, no. 2 (2010): 257–66.

Haase, L., et al. "When the Brain Does Not Adequately Feel the Body: Links Between Low Resilience and Interoception." *Biological Psychology* 113 (2016): 37–45.

Hao, W.Y., and Y. Chen. "Backward Walking Training Improves Balance in School-Aged Boys." *Sports Medicine, Arthroscopy, Rehabilitation, Therapy and Technology* 3 (2011): 24.

Harlow, H.F. "The Formation of Learning Sets." *Psychological Review* 56, no. 1 (1949): 51–65.

Harlow, H.F. "The Nature of Love." *American Psychologist* 13, no. 12 (1958): 673–85.

Harlow, H.F., and R.R. Zimmerman. "Affectional Responses in the Infant Monkey." *Science* 130, no. 3373 (1959): 421–32.

Hebb, D.O. *The Organization of Behavior: A Neuropsychological Theory.* New York: Wiley, 1949.

Herbert, B.M., and O. Pollatos. "The Body in the Mind: On the Relationship Between Interoception and Embodiment." *Topics in Cognitive Science* 4, no. 4 (2012): 692–704.

Hindi, F.S. "How Attention to Interoception Can Inform Dance/Movement Therapy." *American Journal of Dance Therapy* 34 (2012): 129–40.

Huang, R.H., and Y.N. Shih. "Effects of Background Music on Concentration of Workers." *Work* 38, no. 4 (2011): 383–87.

Huron, D. "Is Music an Evolutionary Adaptation?" In *The Cognitive Neuroscience of Music*, edited by I. Peretz and Robert Zatorre, 57–75. New York: Oxford University Press, 2012.

Insel, T.R., and R.D. Fernald. "How the Brain Processes Social Information: Searching for the Social Brain." *Annual Review of Neuroscience* 27 (2004): 697–722.

Jola, C., et al. "Proprioceptive Integration and Body Representation: Insights Into Dancers' Expertise." *Experimental Brain Research* 213, no. 2–3 (2011): 257–65.

Kang, H.J., and V.J. Williamson. "Background Music Can Aid Second Language Learning." *Psychology of Music* 42, no. 5 (2013): 728–47.

Karpati, F.J., et al. "Dance and the Brain: A Review." *Annals of the New York Academy of Sciences* 1337 (2015): 140–46.

Kattenstroth, J.-C., et al. "Six Months of Dance Intervention Enhances Postural, Sensorimotor, and Cognitive Performance in Elderly Without Affecting Cardio-Respiratory Functions." *Frontiers in Aging Neuroscience* 5 (2013): 5.

Kattenstroth, J.-C., et al. "Superior Sensory, Motor, and Cognitive Performance in Elderly Individuals With Multi-Year Dancing Activities." *Frontiers in Aging Neuroscience* 2 (2010): 31.

Keeler, J.R., et al. "The Neurochemistry and Social Flow of Singing: Bonding and Oxytocin." *Frontiers in Human Neuroscience* 9 (2015): 518.

Koelsch, S., et al. "Effects of Music Listening on Cortisol Levels and Propofol Consumption During Spinal Anesthesia." *Frontiers in Psychology* 2 (2011): 58.

Kringelbach, M.L., and K.C. Berridge. "Towards a Functional Neuro-Anatomy of Pleasure and Happiness." *Trends in Cognitive Sciences* 13 (2009): 479–87.

Laland, K., et al. "The Evolution of Dance." *Current Biology* 26, no. 1 (2016): R5–9.

Laukka, P. "Uses of Music and Psychological Well-Being Among the Elderly." *Journal of Happiness Studies* 8 (2007): 215–41.

Lieberwirth, C., and Z. Wang. "Social Bonding: Regulation by Neuropeptides." *Frontiers in Neuroscience* 8, no. 8 (2014): 171.

Lundqvist, L.-O., et al. "Emotional Responses to Music: Experience, Expression, and Physiology." *Psychology of Music* 37, no. 1 (2008): 61–90.

Moradzadeh, L., et al. "Musical Training, Bilingualism, and Executive Function: A Closer Look at Task Switching and Dual-Task Performance." *Cognitive Science* 39, no. 5 (2015): 992–1020.

Morrison, I., et al. "The Skin as a Social Organ." *Experimental Brain Research* 204 (2010): 305–14.

Musacchia, G., et al. "Relationships Between Behavior, Brainstem and Cortical Encoding of Seen and Heard Speech in Musicians and Non-Musicians." *Hearing Research* 241, no. 1–2 (2008): 34–42.

Nelson, A., et al. "The Impact of Music on Hypermetabolism in Critical Illness." *Current Opinion in Clinical Nutrition and Metabolic Care* 11, no. 6 (2008): 790–94.

Netter, F.H., et al. *Atlas of Neuroanatomy and Neurophysiology: Selections From the Netter Collection of Medical Illustrations.* Teterboro, NJ: Icon, 2002.

Pedersen, B.K., and C. Brandt. "The Role of Exercise-Induced Myokines in Muscle Homeostasis and the Defense Against Chronic Diseases." *Journal of Biomedicine and Biotechnology* (2010).

Peretz, I., and R.J. Zatorre. "Brain Organization for Music Processing." *Annual Review of Psychology* 56 (2005): 89–114.

Petrini, K., et al. "Action Expertise Reduces Brain Activity for Audiovisual Matching Actions: An fMRI Study With Expert Drummers." *NeuroImage* 56, no. 3 (2011): 1480–92.

Pollatos, O., et al. "On the Relationship Between Interoceptive Awareness, Emotional Experience, and Brain Processes." *Cognitive Brain Research* 25, no. 3 (2005): 948–62.

Roberts, I.D., and B.M. Way. "Using 'Hug Drugs' to Understand Affiliative Behavior: The Value of the Social Neurochemistry Perspective." *Social Cognitive and Affective Neuroscience* 9, no. 8 (2014): 1053–54.

Salimpoor, V.N., et al. "The Rewarding Aspects of Music Listening Are Related to Degree of Emotional Arousal." *PLOS ONE* 4, no. 10 (2009): e7487.

Schachner, A. "Auditory-Motor Entrainment in Vocal Mimicking Species: Additional Ontogenetic and Phylogenetic Factors." *Communicative and Integrative Biology* 3, no. 3 (2010): 290–93.

Schachner, A., et al. "Spontaneous Motor Entrainment to Music in Multiple Vocal Mimicking Species." *Current Biology* 19, no. 10 (2009): 831–36.

Sejnowski, T. J. "The Book of Hebb." *Neuron* 24, no. 4 (1999): P773–76.

Seth, A.K. "Interoceptive Inference, Emotion, and the Embodied Self." *Trends in Cognitive Sciences* 17, no. 11 (2013): 565–73.

Shafir, T., et al. "Emotion Regulation Through Execution, Observation, and Imagery of Emotional Movements." *Brain and Cognition* 82, no. 2 (2013): 219–27.

Shafir, T., et al. "Emotion Regulation Through Movement: Unique Sets of Movement Characteristics Are Associated With and Enhance Basic Emotions." *Frontiers in Psychology* 6 (2016): 2030.

Shafritz, K.M., et al. "The Interaction of Emotional and Cognitive Neural Systems in Emotionally Guided Response Inhibition." *NeuroImage* 31, no. 1 (2006): 468–75.

Shih, Y.N., et al. "Background Music: Effects on Attention Performance." *Work* 42, no. 4 (2012): 573–78.

Silverman, M.J., Baker, F.A., and R.A.R. MacDonald. "Flow and Meaningfulness as Predictors of Therapeutic Outcome Within Songwriting Interventions." *Psychology of Music* 44, no. 6 (2016): 1331–45.

Srhoj, L., et al. "Impact of Motor Abilities on Belly Dance Performance in Female High School Students." *Collegium Antropologicum* 32, no. 3 (2008): 835–41.

Stewart, L., et al. "Brain Changes After Learning to Read and Play Music." *NeuroImage* 20, no. 1 (2003): 71–83.

Stuckey, H.L., and J. Nobel. "The Connection Between Art, Healing, and Public Health: A Review of Current Literature." *American Journal of Public Health* 100, no. 2 (2010): 254–63.

Viggiano, D., et al. "The Kinematic Control During the Backward Gait and Knee Proprioception: Insights From Lesions of the Anterior Cruciate Ligament." *Journal of Human Kinetics* 41, no. 1 (2014): 51–57.

Weng, H.Y., et al. "Compassion Training Alters Altruism and Neural Responses to Suffering." *Psychological Science* 24, no. 7 (2013): 1171–80.

Wiens, S. "Interoception in Emotional Experience." *Current Opinion in Neurology* 18, no. 4 (2005): 442–47.

Wong, P.C.M., et al. "Musical Experience Shapes Human Brainstem Encoding of Linguistic Pitch Patterns." *Nature Neuroscience* 10, no. 4 (2007): 420–22.

Woodward, A.L. "Infants' Ability to Distinguish Between Purposeful and Non-Purposeful Behaviors." *Infant Behavior and Development* 22, no. 2 (1999): 145–60.

Yuan, H., and S.D. Silberstein. "Vagus Nerve and Vagus Nerve Stimulation, a Comprehensive Review: Part II." *Headache* 56, no. 2 (2016): 259–66.

CHAPTER 5

Arcelus, J., et al. "Prevalence of Eating Disorders Amongst Dancers: A Systemic Review and Meta-Analysis." *European Eating Disorders Review* 22, no. 2 (2014): 92–101.

Belardinelli, R., et al. "Waltz Dancing in Patients With Chronic Heart Failure: New Form of Exercise Training." *Circulation: Heart Failure* 1, no. 2 (2008): 107–14.

Bernardi, L., et al. "Dynamic Interactions Between Musical, Cardio-Vascular, and Cerebral Rhythms in Humans." *Circulation* 119, no. 25 (2009): 3171–80.

Bernardi, L., et al. "Cardiovascular, Cerebrovascular, and Respiratory Changes Induced by Different Types of Music in Musicians and Nonmusicians: The Importance of Silence." *Heart* 92 (2006): 445–52.

Blanksby, B.A., and P.W. Reidy. "Heart Rate and Estimated Energy Expenditure During Ballroom Dancing." *British Journal of Sports Medicine* 22, no. 2 (1988): 57–60.

Borstel, J.v. *Herzrasen kann man nicht mähen: Alles über unser wichtigstes Organ.* Berlin: Ullstein Verlag, 2016.

Clarendon, D., and K. Meola. "*Dancing With the Stars* Celebrities Who Lost Weight on the Show: Before and After," *Us Weekly*, October 22, 2019.

Codrons, E., et al. "Spontaneous Group Synchronization of Movements and Respiratory Rhythms." PLOS ONE 9 (2014).

Cugusi, L., et al. "A New Type of Physical Activity From an Ancient Tradition: The Sardinian Folk Dance 'Ballu Sardu.'" *Journal of Dance Medicine and Science* 19, no. 3 (2015): 118–23.

Domene, P.A. et al. "The Health Enhancing Efficacy of Zumba Fitness: An 8-Week Randomised Controlled Study." *Journal of Sports Studies* 34, no. 15 (2016): 1396–404.

Fragala, M.S., et al. "Neuroendocrine-Immune Interactions and Responses to Exercise." *Sports Medicine* 41 (2011): 621–39.

Gildea, J.E., et al. "Trunk Dynamics Are Impaired in Ballet Dancers With Back Pain but Improve With Imagery." *Medicine and Science in Sports and Exercise* 47, no. 8 (2015): 1665–71.

Gurmán-García, A. "Introducing a Latin Ballroom Dance Class to People With Dementia Living in Care Homes, Benefits and Concerns: A Pilot Study." *Dementia* 12, no. 5 (2013): 523–35.

Hanna, J.L. *Dancing for Health: Conquering and Preventing Stress*. New York: AltaMira Press, 2006.

Hogg, J., et al. "An After-School Dance and Lifestyle Education Program Reduces Risk Factors for Heart Disease and Diabetes in Elementary School Children." *Journal of Pediatric Endocrinology and Metabolism* 25, no. 5–6 (2014): 509–16.

Hui, E., et al. "Effects of Dance on Physical and Psychological Well-Being in Older Persons." *Archives of Gerontology and Geriatrics* 49, no. 1 (2009): e45–50.

Karin, J. "Recontextualizing Dance Skills: Overcoming Impediments to Motor Learning and Expressivity in Ballet Dancers." *Frontiers in Psychology* 7 (2016): 431.

Karin, J., et al. "Mental Training." In *Dancer Wellness*, edited by V. Wilmerding and D. Krasnow. Champaign, IL: Human Kinetics, 2016.

King, D.E., et al. "Inflammatory Markers and Exercise: Differences Related to Exercise Type." *Medicine and Science in Sports and Exercise* 35, no. 4 (2003): 575–81.

Krampe, J., et al. "Does Dance-Based Therapy Increase Gait Speed in Older Adults With Chronic Lower Extremity Pain: A Feasibility Study." *Geriatric Nursing* 35, no. 5 (2014): 339–44.

Mangeri, F., et al. "A Standard Ballroom and Latin Dance Program to Improve Fitness and Adherence to Physical Activity in Individuals With Type 2 Diabetes and in Obesity." *Diabetology Metabolism Syndrome* 6 (2014): 74.

Merom, D., et al. "Dancing Participation and Cardiovascular Disease Mortality: A Pooled Analysis of 11 Population-Based British Cohorts." *American Journal of Preventative Medicine* 50, no. 6 (2016): 756–60.

Narici, M., et al. "Does Dancing in Old Age Afford Neuromuscular Protection?" *Proceedings of the Physiological Society* 35 (2016): PC31.

National Institute of Neurological Disorders and Stroke. "Low Back Pain Fact Sheet." March 2020, https://www.ninds.nih.gov/Disorders/Patient-Caregiver-Education/Fact-Sheets/Low-Back-Pain-Fact-Sheet.

Nogueira, R.C., et al. "An In-School Exercise Intervention to Enhance Bone and Reduce Fat in Girls: The Capo Kids Trial." *Bone* 68 (2014): 92–99.

Porges, S.W. "Love: An Emergent Property of the Mammalian Autonomic Nervous System." *Psychoneuroendocrinology* 23, no. 8 (1998): 837–61.

Porges, S.W. "The Polyvagal Perspective." *Biological Psychology* 74, no. 2 (2007): 116–43.

Richardson, J.D., et al. "The Dancing Heart." *European Heart Journal: Acute Cardiovascular Care* 5, no. 1 (2016): 96–97.

Schmitt-Sody, M., et al. "Rehabilitation and Sport Following Total Hip Replacement." *Orthopade* 40, no. 6 (2011): 513–19.

Serrano-Guzmán, M., et al. "Effectiveness of a Flamenco and Sevillanas Program to Enhance Mobility, Balance, Physical

Activity, Blood Pressure, Body Mass, and Quality of Life in Postmenopausal Women Living in the Community in Spain: A Randomized Clinical Trial." *Menopause* 23, no. 9 (2016): 965-73.

Sharma, A., et al. "Exercise for Mental Health." *The Primary Care Companion to the Journal of Clinical Psychiatry* 8, no. 2 (2006): 106.

Tilley, A.J., and P. Bohle. "Twisting the Night Away: The Effects of All-Night Disco Dancing on Reaction Time." *Perceptual and Motor Skills* 66, no. 1 (1988): 107-12.

Tsakiris, M. and H. Critchley. "Interoception Beyond Homeostasis: Affect, Cognition and Mental Health." *Philosophical Transactions of the Royal Society of London: Series B* 371, no. 1708 (2016).

Wigaeus, E., and A. Kilborn. "Physical Demands During Folk Dancing." *European Journal of Applied Physiology and Occupational Physiology* 45 (1980): 177-83.

Williford, H.N., M. Scharff-Olson, and D.L. Blessing. "The Physiological Effects of Aerobic Dance: A Review." *Sports Medicine* 8, no. 6 (1989): 335-45.

Zimmerman, S., et al. "Systematic Review: Effective Characteristics of Nursing Homes and Other Residential Long-Term Care Settings for People With Dementia." *Journal of the American Geriatric Society* 61, no. 8 (2013): 1399-409.

CHAPTER 6

Aktas, G., and F. Ogce. "Dance as a Therapy for Cancer Prevention." *Asian Pacific Journal of Cancer Prevention* 6, no. 3 (2005): 408-11.

Bernstein, B. "Dancing Beyond Trauma: Women Survivors of Sexual Abuse." In *Dance and Other Expressive Art Therapies: When Words Are Not Enough*, edited by F.J. Levy, J. Fried, and F. Leventhal, 41-58. New York: Routledge, 1995.

Berridge, K.C., and M.L. Kringelbach. "Affective Neuroscience of Pleasure: Reward in Humans and Animals." *Psychopharmacology* 199, no. 3 (2008): 457–80.

Berridge, K.C., and M.L. Kringelbach. "Pleasure Systems in the Brain." *Neuron* 86, no. 3 (2015): 646–64.

Dedovic, K., et al. "The Brain and the Stress Axis: The Neural Correlates of Cortisol Regulation in Response to Stress." *NeuroImage* 47, no. 3 (2009): 864–71.

Förster, J., and L. Werth. "Regulatory Focus: Classic Findings and New Directions." In *The Psychology of Goals*, edited by G.B. Moskowitz and H. Grant, 392–420. New York: Guildford Press, 2009.

Garrido, S., and E. Schubert. "Adaptive and Maladaptive Attraction to Negative Emotions in Music." *Musicae Scientiae* 17, no. 2 (2013): 147–66.

Gray, A.E. "Dance/Movement Therapy With Refugee and Survivor Children: A Healing Pathway Is a Creative Process." In *Creative Interventions With Traumatized Children*, 2nd ed., edited by Cathy A. Malchiodi, 169–90. New York: Guildford Press, 2015.

Gray, M. J., and T.W. Lombardo. "Complexity of Trauma Narratives as an Index of Fragmented Memory in PTSD: A Critical Analysis." *Applied Cognitive Psychology* 15, no. 7 (2001): S171–86.

Hanna, J. L. "The Power of Dance: Health and Healing." *Journal of Alternative and Complementary Medicine* 1, no. 4 (1995): 323–31.

Heiberger, L., et al. "Impact of a Weekly Dance Class on the Functional Mobility and on the Quality of Life of Individuals With Parkinson's Disease." *Frontiers in Aging Neuroscience* 3 (2011): 14.

Hermann, J. *Die Narben der Gewalt: Traumatische Erfahrungen verstehen und überwinden* [Trauma and Recovery: The Aftermath of Violence—From Domestic Abuse to Political Terror] Paderborn, DE: Junfermann, 2006.

Hogue, J.D., et al. "'So Sad and Slow, So Why Can't I Turn Off the Radio': The Effects of Gender, Depression, and Absorption on Liking Music That Induces Sadness and Music That Induces Happiness." *Psychology of Music* 44, no. 4 (2016): 816–29.

Huron, D. "Why Is Sad Music Pleasurable? A Possible Role for Prolactin." *Musicae Scientiae* 15, no. 2 (2011): 146–58.

Jelinek, L., et al. "The Organization of Autobiographical and Non-autobiographical Memory in Posttraumatic Stress Disorder (PTSD)." *Journal of Abnormal Psychology* 118, no. 2 (2009): 288–98.

Jeong, Y.J., et al. "Dance Movement Therapy Improves Emotional Responses and Modulates Neurohormones in Adolescents With Mild Depression." *International Journal of Neuroscience* 115, no. 12 (2005): 1711–20.

Kiepe, M.S., et al. "Effects of Dance Therapy and Ballroom Dances on Physical and Mental Illnesses: A Systematic Review." *Arts in Psychotherapy* 39, no. 5 (2012): 404–11.

Kirk, A.E. *Dance/Movement Therapy for Adult Women With Post Traumatic Stress Disorder: A QuasiExperimental Study of Symptom Reduction and Integration*, Dissertation Abstracts International Section A: Humanities and Social Sciences (2015).

Koch, S., et al. "Effects of Dance Movement Therapy and Dance on Health-Related Psychological Outcomes: A Meta-Analysis." *Arts in Psychotherapy* 41, no. 1 (2014): 46–64.

Krantz, A. "Growing Into Her Body: Dance/Movement Therapy for Women With Eating Disorders." *American Journal of Dance Therapy* 21, no. 2 (1999): 81–103.

Kreutz, G., et al. "Psychoneuroendocrine Research on Music and Health: An Overview." In *Music, Health, and Wellbeing*, edited by R. MacDonald, G. Kruetz, and L. Mitchell, 457–76. Oxford: Oxford University Press, 2012.

Kringelbach, M.L., and K.C. Berridge. "Towards a Functional Neuro-Anatomy of Pleasure and Happiness." *Trends in Cognitive Sciences* 13, no. 11 (2009): 479–87.

LeDoux, J. "The Amygdala." *Current Biology* 17, no. 20 (2007): R868–74.

LeDoux, J. "The Emotional Brain, Fear, and the Amygdala." *Cellular and Molecular Neurobiology* 23, no. 4–5 (2003): 727–38.

LeDoux, J. "Rethinking the Emotional Brain." *Neuron* 73, no. 4 (2012): 653–76.

Lee, T.C., et al. "Dance/Movement Therapy for Children Suffering From Earthquake Trauma in Taiwan: A Preliminary Exploration." *Arts in Psychotherapy* 40, no. 1 (2013): 151–57.

Leknes, S., and I. Tracey. "A Common Neurobiology for Pain and Pleasure." *Nature Reviews Neuroscience* 9 (2008): 314–20.

Levy, B.J., and M.C. Anderson. "Individual Differences in the Suppression of Unwanted Memories: The Executive Deficit Hypothesis." *Acta Psychologica* 127, no. 3 (2008): 623–35.

Lima, R. "Balance Assessment in Deaf Children and Teenagers Prior to and Post Capoeira Practice Through the Berg Balance Scale." *International Tinnitus Journal* 21, no. 2 (2017): 77–82.

Margariti, A. "Review of Dancing for Health." *Body, Movement and Dance in Psychotherapy* 6, no. 1 (2011): 77–80.

McCarty, R. "Fight-or-Flight Response." In *Stress: Concepts, Cognition, Emotion and Behavior*, edited by G. Fink, 33–37. Academic Press, 2016.

Meekums, B., et al. "Dance Movement Therapy for Depression." *Cochrane Database of Systematic Reviews* 2 (2015).

Milner, P.M. "Brain-Stimulation Reward: A Review." *Canadian Journal of Psychology* 45, no. 1 (1991): 1–36.

Mori, K., and M. Iwanaga. "Pleasure Generated by Sadness: Effect of Sad Lyrics on the Emotions Induced by Happy Music." *Psychology of Music* 42, no. 5 (2014): 643–52.

Murcia, C.Q., and G. Kreutz. "Dance and Health: Exploring Interactions and Implications." In *Music, Health, and Wellbeing*, edited by R. MacDonald, G. Kreutz, and L. Mitchell, 125-35. Oxford: Oxford University Press, 2012.

Olds, J., and P. Milner. "Positive Reinforcement Produced by Electrical Stimulation of Septal and Other Regions of the Rat Brain." *Journal of Comparative and Physiological Psychology* 47, no. 6 (1954): 419-27.

Ottaviani, C., et al. "Cognitive, Behavioral, and Autonomic Correlates of Mind Wandering and Perseverative Cognition in Major Depression." *Frontiers in Neuroscience* 8 (2015): 1-9.

Pinniger, R., et al. "Argentine Tango Dance Compared to Mindfulness Meditation and a Waiting-List Control: A Randomised Trial for Treating Depression." *Complementary Therapies in Medicine* 20, no. 6 (2012): 377-84.

Pratt, R.R. "Art, Dance, and Music Therapy." *Physical Medicine and Rehabilitation Clinics of North America* 15, no. 4 (2004): 827-41.

Quiroga Murcia, C., et al. "Emotional and Neurohumoral Responses to Dancing Tango Argentino: The Effects of Music and Partner." *Music and Medicine* 1, no. 1 (2009): 14-21.

Quiroga Murcia, C., et al. "Shall We Dance? An Exploration of the Perceived Benefits of Dancing on Well-Being." *Arts & Health* 2, no. 2 (2010): 149-63.

Ramaprasad, D. "Emotions: An Indian Perspective." *Indian Journal of Psychiatry* 55, suppl. 2 (2013): S153-56.

Ritter, M., and K.G. Low. "Effects of Dance/Movement Therapy: A Meta-Analysis." *Arts in Psychotherapy* 10 (1996): 1806.

Röhricht, F. "Body Oriented Psychotherapy. The State of the Art in Empirical Research and Evidence-Based Practice: A Clinical Perspective." *Body, Movement and Dance in Psychotherapy* 4, no. 2 (2009): 135-56.

Sachs, M.E., et al. "The Pleasures of Sad Music: A Systematic Review." *Frontiers in Human Neuroscience* 9 (2015): 404.

Selman, L.E., et al. "A Mixed-Methods Evaluation of Complementary Therapy Services in Palliative Care: Yoga and Dance Therapy." *European Journal of Cancer Care* 21, no. 1 (2012): 87–97.

Shafir, T. "Using Movement to Regulate Emotion: Neurophysiological Findings and Their Application in Psychotherapy." *Frontiers in Psychology* 7 (2016): 1451.

Shafir, T., et al. "Emotion Regulation Through Execution, Observation, and Imagery of Emotional Movements." *Brain and Cognition* 82, no. 2 (2013): 219–27.

Shafir, T., et al. "Emotion Regulation Through Movement: Unique Sets of Movement Characteristics Are Associated With and Enhance Basic Emotions." *Frontiers in Psychology* 6 (2016): 2030.

Spors, M. "Eating Disorders in Women With a History of Abuse and Dance/Movement Therapy as a Treatment Modality." PhD diss., Universität Göttingen, 1997.

Stone, N. "*Dancing With the Stars*' First-Ever Blind Contestant Wows Judges With Tear-Jerking Performance." *People*, September 24, 2018.

Taruffi, L. and S. Koelsch. "The Paradox of Music-Evoked Sadness: An Online Survey." PLOS ONE 9, no. 10 (2014).

Torres, E.B. "Commentary On: An Exploration of Sensory and Movement Differences From the Perspective of Individuals With Autism." *Frontiers in Integrative Neuroscience* 9 (2015): 20.

Torres, E.B., et al. "Autism: The Micro-Movement Perspective." *Frontiers in Integrative Neuroscience* 7 (2013): 32.

Torres, E.B., and A.M. Donnellan. "Editorial for Research Topic 'Autism: The Movement Perspective.'" *Frontiers in Integrative Neuroscience* 9 (2015): 12.

Torres-McGehee, T.M., et al. "Body Image, Anthropometric Measures, and Eating-Disorder Prevalence in Auxiliary Unit Members." *Journal of Athletic Training* 44, no. 4 (2009): 418–26.

Tsimaras, V.K., et al. "The Effect of a Traditional Dance Training Program on the Physical Fitness of Adults With Hearing Loss." *Journal of Strength and Conditioning Research* 24, no. 4 (2010): 1052–58.

Vinesett, A.L., et al. "Modified African Ngoma Healing Ceremony for Stress Reduction: A Pilot Study." *Journal of Alternative Complementary Medicine* 23, no. 10 (2017): 800–804.

West, J., et al. "Effects of Hatha Yoga and African Dance on Perceived Stress, Affect, and Salivary Cortisol." *Annals of Behavioural Medicine* 28, no. 2 (2004): 114–18.

Woolf, S., and P. Fisher. "The Role of Dance Movement Psychotherapy for Expression and Integration of the Self in Palliative Care." *International Journal of Palliative Nursing* 21, no. 7 (2015): 340–48.

Wu, D., et al. "A Biomarker Characterizing Neurodevelopment With Applications in Autism." *Scientific Reports* 8 (2018).

CHAPTER 7

Alpert et al. "The Effect of Modified Jazz Dance on Balance, Cognition, and Mood in Older Adults." *Journal of the American Academy of Nurse Practice* 21, no. 2 (2009): 108–15.

American Speech-Language-Hearing Association. "Dementia." https://www.asha.org/PRPSpecificTopic.aspx?folderid= 8589935289&section=Incidence_and_Prevalence.

An, S.-Y., et al. "Effect of Belly Dancing on Urinary Incontinence-Related Muscles and Vaginal Pressure in Middle-Aged Women." *Journal of Physical Therapy Science* 29, no. 3 (2017): 384–86.

Burzynska, A.Z., et al. "The Dancing Brain: Structural and Functional Signatures of Expert Dance Training." *Frontiers in Human Neuroscience* 11 (2017): 566.

Butt, C.A. "'Move Your Arm Like a Swan': Dance for PD Demedicalizes Parkinson Disease." JAMA 317, no. 4 (2017): 342–43.

Cepeda, C., et al. "Effect of an Eight-Week Ballroom Dancing Program on Muscle Architecture in Older Adult Females." *Journal of Aging and Physical Activity* 23, no. 4 (2015): 607–12.

Coubard, O., et al. "Practice of Contemporary Dance Improves Cognitive Flexibility in Aging." *Frontiers in Aging Neuroscience* 3 (2011): 13.

Cruz-Ferreira, A., et al. "Creative Dance Improves Physical Fitness and Life Satisfaction in Older Women." *Research on Aging* 37, no. 8 (2015): 837–55.

Foster, E.R., et al. "A Community-Based Argentine Tango Dance Program Is Associated With Increased Activity Participation Among Individuals With Parkinson's Disease." *Archives of Physical Medicine and Rehabilitation* 94, no. 2 (2013): 240–49.

Foster, N.A., and E.R Valentine. "The Effect of Auditory Stimulation on Autobiographical Recall in Dementia." *Experimental Aging Research* 27, no. 3 (2001): 215–28.

Girault, J.A., and P. Greengard. "The Neurobiology of Dopamine Signaling." *Archives of Neurology* 61, no. 5 (2004): 641–44.

Gomes de la Silva Borges, E., et al. "The Effect of Ballroom Dance on Balance and Functional Autonomy Among the Isolated Elderly." *Archives of Gerontology and Geriatrics* 55, no. 2 (2012): 492–96.

Granacher, U., et al. "Effects of a Salsa Dance Training on Balance and Strength Performance in Older Adults." *Gerontology* 58, no. 4 (2012): 305–12.

Hackney, M.E., and G.M. Earhart. "Effects of Dance on Balance and Gait in Severe Parkinson Disease: A Case Study." *Disability and Rehabilitation* 32, no. 8 (2010): 679–84.

Hackney, M.E., and G.M. Earhart. "Short Duration, Intensive Tango Dancing for Parkinson Disease: An Uncontrolled Pilot Study." *Complementary Therapies in Medicine* 17, no. 4 (2009) 203–7.

Hwang, P.W., and K.L. Braun. "The Effectiveness of Dance Interventions to Improve Older Adults' Health: A Systematic Literature Review." *Alternative Therapies in Health and Medicine* 21, no. 5 (2015): 64–70.

Jacobsen, J.H., et al. "Why Musical Memory Can Be Preserved in Advanced Alzheimer's Disease." *Brain* 138, no. 8 (2015): 2438–50.

Janyacharoen, T., et al. "Physical Performance in Recently Aged Adults After 6 Weeks Traditional Thai Dance: A Randomized Controlled Trial." *Clinical Intervention in Aging* 8 (2013): 855–59.

Julian, A.M., and M. Paisley. "Towards the Design, Development, and Evaluation of a Personalised Music Intervention in Dementia." *Alzheimer's & Dementia* 10, no. 4 (2014): P3–379.

Koch, S.C., et al. "The Embodied Self in Parkinson's Disease: Feasibility of a Single Tango Intervention for Assessing Changes in Psychological Health Outcomes and Aesthetic Experience." *Frontiers in Neuroscience* 10 (2016): 287.

Koger, S.M., and M. Brotons. "Music Therapy for Dementia Symptoms." *Cochrane Database of Systematic Reviews* 2 (2000).

Lötzke, D., et al. "Argentine Tango in Parkinson Disease: A Systematic Review and Meta-Analysis." BMC *Neurology* 15, no. 1 (2015): 226.

Marmeleira, J. "An Examination of the Mechanisms Underlying the Effects of Physical Activity on Brain and Cognition." *European Review of Aging and Physical Activity* 10 (2013): 83–94.

Marras, C., et al. "Prevalence of Parkinson's Disease Across North America." *npj Parkinson's Disease* 4, no. 1 (2018): 1–7.

McKee, K.E., et al. "The Effects of Adapted Tango on Spatial Cognition and Disease Severity in Parkinson's Disease." *Journal of Motor Behaviour* 45, no. 6 (2013): 519–29.

McKinley, P., et al. "Effect of a Community-Based Argentine Tango Dance Program on Functional Balance and Confidence in Older Adults." *Journal of Aging and Physical Activity* 16, no. 4 (2008): 435–53.

McNeely, M.E., et al. "Impacts of Dance on Non-Motor Symptoms, Participation, and Quality of Life in Parkinson Disease and Healthy Older Adults." *Maturitas* 82, no. 4 (2015): 336–41.

McNeely, M.E., et al. "Differential Effects of Tango Versus Dance for PD in Parkinson Disease." *Frontiers in Aging Neuroscience* 7 (2015): 239.

Müller, P., et al. "Evolution of Neuroplasticity in Response to Physical Activity in Old Age: The Case for Dancing." *Frontiers in Aging Neuroscience* 9 (2017): 56.

Müller, P., et al. "Präventionsstrategien gegen Demenz." *Zeitschrift für Gerontologie und Geriatrie* 50 (2017): 89–95.

Myers, N. "Dance Your PhD: Embodied Animations, Body Experiments, and the Affective Entanglements of Life Science Research." *Body & Society* 18, no. 1 (2012): 151–89.

Prince, M., et al. "The Global Prevalence of Dementia: A Systematic Review and Metaanalysis." *Alzheimer's and Dementia* 9, no. 1 (2013): 63–75.

Rehfeld, K., et al. "Dancing or Fitness Sport? The Effects of Two Training Programs on Hippocampal Plasticity and Balance Abilities in Healthy Seniors." *Frontiers in Human Neuroscience* 11 (2017): 305.

Salmon, D.P., and M.W. Bondi. "Neuropsychological Assessment of Dementia." *Annual Review of Psychology* 60 (2009): 257–82.

Shigematsu, R., et al. "Dance-Based Aerobic Exercise May Improve Indices of Falling Risk in Older Women." *Age and Aging* 31 (2002): 261–66.

"Study Is Nation's First to Examine Prevalence of Dance Among Youth." Arnold School of Public Health, University of South Carolina, October 26, 2011, http://www.asph.sc.edu/news/exsc_dance.htm.

Valenzuela, M.J., and P. Sachdev. "Brain Reserve and Dementia: A Systematic Review." *Psychological Medicine* 36, no. 4 (2006): 441–54.

Verghese, J., et al. "Leisure Activities and the Risk of Dementia in the Elderly." *New England Journal of Medicine* 348, no. 25 (2003): 2508–16.

Wallmann H.W., et al. "The Effect of a Senior Jazz Dance Class on Static Balance in Healthy Women Over 50 Years of Age: A Pilot Study." *Biological Research for Nursing* 10, no. 3 (2009): 257–66.

Westheimer, O. "Dance and Parkinson's Disease." *Movement Disorders* (2010).

Westheimer, O. "Why Dance for Parkinson's Disease." *Topics in Geriatric Rehabilitation* 24, no. 2 (2008): 127–40.

Westheimer, O., et al. "Dance for PD: A Preliminary Investigation of Effects on Motor Function and Quality of Life Among Persons With Parkinson's Disease (PD)." *Journal of Neural Transmission* 122, no. 9 (2015): 1263–70.

World Health Organization. "Dementia: A Public Health Priority." *Dementia*. (2012).

World Health Organization. *Global Recommendation on Physical Activity for Health*. (2004).

World Health Organization. "Infographic on Dementia." September 20, 2017, https://www.who.int/mental_health/neurology/dementia/infographic_dementia/en/.

World Health Organization. "Physical Activity." February 23, 2018, https://www.who.int/en/news-room/fact-sheets/detail/physical-activity.

CHAPTER 8

Aronoff, J. "How We Recognize Angry and Happy Emotion in People, Places, and Things." *Cross-Cultural Research* 40, no. 1 (2006): 83–105.

Bar, M., and M. Neta. "Humans Prefer Curved Visual Objects." *Psychological Science* 17, no. 8 (2006): 645–48.

Bar, M., and M. Neta. "Visual Elements of Subjective Preference Modulate Amygdala Activation." *Neuropsychologia* 45, no. 10 (2007): 2191–200.

Bennassar, M. and R. de Ayreflor. *Influencia del Baile Popular Mallorquin en la Pintura de Coll Bardolet.* Palma: Santuari de Lluc, 1997.

Bonny, J., et al. "Hip Hop Dance Experience Linked to Sociocognitive Ability." PLOS ONE 12, no. 2 (2017).

Boing, L. et al. "Benefits of Belly Dance on Quality of Life, Fatigue, and Depressive Symptoms in Women With Breast Cancer: A Pilot Study of a Non-Randomised Clinical Trial." *Journal of Body and Movement Therapy* 22, no. 2 (2018): 460–66.

Calvo-Merino, B., et al. "Action Observation and Acquired Motor Skills: An fMRI Study With Expert Dancers." *Cerebral Cortex* 15, no. 8 (2005): 1243–49.

Calvo-Merino, B., et al. "Seeing or Doing? Influence of Visual and Motor Familiarity in Action Observation." *Current Biology* 16, no. 9 (2006): 1905–10.

Chatterjee, A. "Prospects for a Cognitive Neuroscience of Visual Aesthetics." *Bulletin of Psychology and the Arts* 4 (2003): 55–60.

Christensen, J.F., and B. Calvo-Merino. "Dance as a Subject for Empirical Aesthetics." *Psychology of Aesthetics, Creativity, and the Arts* 7, no. 1 (2013): 76–88.

Christensen, J.F., et al. "Affective Responses to Dance." *Acta Psychologica* 168 (2016): 91–105.

Clignet, R. "Review of *Dance, Sex and Gender: Signs of Identity, Dominance, Defiance, and Desire* by Judith Lynne Hanna." *Contemporary Sociology* 18, no. 4 (1989): 602–3.

Cross, E.S., et al. "A Review and Critical Analysis of How Cognitive Neuroscientific Investigations Using Dance Can Contribute to Sport Psychology." *International Review of Sport and Exercise Psychology* 7, no. 1 (2013): 42–71.

Cross, E.S., et al. "Building a Motor Simulation De Novo: Observation of Dance by Dancers." *NeuroImage* 31, no. 3 (2006): 1257–67.

Cross, E.S., et al. "The Impact of Aesthetic Evaluation and Physical Ability on Dance Perception." *Frontiers in Human Neuroscience* 5 (2011): 102.

Cross, E.S., and L.F. Ticini. "Neuroaesthetics and Beyond: New Horizons in Applying the Science of the Brain to the Art of Dance." *Phenomenology and the Cognitive Sciences* 11 (2011): 5–16.

Daprati, E., et al. "A Dance to the Music of Time: Aesthetically-Relevant Changes in Body Posture in Performing Art." PLOS ONE 4, no. 3 (2009).

De Warren, R. *Destiny's Waltz: In Step With Giants*. New York: Strategic Book Publishing & Rights Agency, 2009.

Dimler, A.J., et al. "'I Kinda Feel Like Wonder Woman': An Interpretative Phenomenological Analysis of Pole Fitness and Positive Body Image." *Journal of Sport and Exercise Psychology* 39, no. 5 (2017): 1–13.

Dutton, D.G., and A.P. Aron. "Some Evidence for Heightened Sexual Attraction Under Conditions of High Anxiety." *Journal of Personality and Social Psychology* 30, no. 4 (1974): 510–17.

Ferguson, S., et al. "Movement in a Contemporary Dance Work and Its Relation to Continuous Emotional Response." Proceedings of the International Conference on New Interfaces for Musical Expression, Sydney, Australia, 2010.

Foster, S.L., ed. *Worlding Dance—Studies in International Performance*. London: Palgrave Macmillan, 2009.

Hanna, J.L. "Dance and Sexuality: Many Moves." *Journal of Sex Research* 47, no. 2 (2010): 212–41.

Hanna, J.L. "The Power of Dance: Health and Healing." *Journal of Alternative and Complementary Medicine* 1, no. 4 (1995): 323–31.

Jang, S.H., and F.E. Pollick. "Experience Influences Brain Mechanisms of Watching Dance." *Dance Research* 29 (2011): 352–77.

Jola, C., et al. "Motor Simulation Without Motor Expertise: Enhanced Corticospinal Excitability in Visually Experienced Dance Spectators." PLOS ONE 7, no. 3 (2012).

Jola, C., et al. "The Experience of Watching Dance: Phenomenological-Neuroscience Duets." *Phenomenology and the Cognitive Sciences* 11 (2011): 17–37.

Jola, C., et al. "Arousal Decrease in Sleeping Beauty: Audiences' Neurophysiological Correlates to Watching a Narrative Dance Performance of Two-and-a-Half Hours." *Dance Research* 29 (2011): 378–403.

Kaltsatou, A., et al. "Physical and Psychological Benefits of a 24-Week Traditional Dance Program in Breast Cancer Survivors." *Journal of Body and Movement Therapy* 15, no. 2 (2011): 162–67.

Khorsandi, S. *The Art of Persian Dance: Shahrzad Technique.* Richmond, CA: Shahrzad Dance Academy, 2015.

Kim, S., and J. Kim. "Mood After Various Brief Exercise and Sport Modes: Aerobic, Hip Hop Dancing, Ice Skate and Body Conditioning." *Perceptual Motor Skills* 104, no. 3 (2007): 1265–70.

Kirsch, L.P., et al. "Dance Experience Sculpts Aesthetic Perception and Related Brain Circuits." *Annals of the New York Academy of Sciences* 1337, no. 1 (2015): 130–39.

Kirsch, L.P., et al. "The Impact of Sensorimotor Experience on Affective Evaluation of Dance." *Frontiers in Human Neuroscience* 7 (2013): 521.

Kirsch, L.P., et al. "Shaping and Reshaping the Aesthetic Brain: Emerging Perspectives on the Neurobiology of Embodied Aesthetics." *Neuroscience and Biobehavioral Reviews* 62 (2016): 56–68.

Kuroda, Y., et al. "Stress, Emotions, and Motivational States Among Traditional Dancers in New Zealand and Japan." *Psychological Reports* 120, no. 5 (2017): 895–913.

Larson, C.L., et al. "The Shape of Threat: Simple Geometric Forms Evoke Rapid and Sustained Capture of Attention." *Emotion* 7, no. 3 (2007): 526-34.

Lovatt, P. *Dance Psychology*. Norfolk, UK: Lulu Publishers, 2018.

Maraz, A., et al. "Why Do You Dance? Development of the Dance Motivation Inventory (DMI)." *PLOS ONE* 10, no. 3 (2015).

Maraz, A., et al. "An Empirical Investigation of Dance Addiction." *PLOS ONE* 10, no. 5 (2015).

Moreira, S.R., et al. "Ten Weeks of Capoeira Progressive Training Improved Cardiovascular Parameters in Male Practitioners." *Journal of Sports Medicine and Physical Fitness* 57, no. 3 (2017): 289-98.

Murrock, C., et al. "A Culturally-Specific Dance Intervention to Increase Functional Capacity in African American Women." *Journal of Cultural Diversity* 15, no. 4 (2008): 168-73.

Nawrocka, A., et al. "Effects of Exercise Training Experience on Hand Grip Strength, Body Composition and Postural Stability in Fitness Pole Dancers." *Journal of Sports Medicine and Physical Fitness* 57, no. 9 (2017): 1098-1103.

Ramón y Cajal, S. "The Structure and Connexions of Neurons." Nobel Lecture, December 12, 1906, https://www.nobelprize.org/prizes/medicine/1906/cajal/lecture/.

Robinson, C., et al. "A Review of Hip Hop-Based Interventions for Health Literacy, Health Behaviors, and Mental Health." *Journal of Racial and Ethnic Health Disparities* 5, no. 3 (2018): 468-84.

Shay, A. *Choreophobia—Solo Improvised Dance in the Iranian World*. Costa Mesa, CA: Mazda, 1999.

Stevens, C., and S. McKechnie. "Thinking in Action: Thought Made Visible in Contemporary Dance." *Cognitive Processing* 6, no. 4 (2005): 243-52.

Stevens, C.J., et al. "Cognition and the Temporal Arts: Investigating Audience Response to Dance Using PDAs That Record Continuous Data During Live Performance." *International Journal of Human-Computer Studies* 67, no. 9 (2009): 800–13.

Zeki, S. "Artistic Creativity and the Brain." *Science* 293, no. 5527 (2001): 51–52.

INDEX OF HEALTH CONCERNS AND DANCES

abuse, 143–44, 220
adrenaline, 44, 154
adumu (Maasai dance), 85
African dance, 85, 88, 158, 198, 244–45
aging, 178–85
amusia (music blindness), 37
anxiety, 160–64
  See also stress
arthrosis, 132–33
attention-deficit hyperactivity disorder (ADHD), 104
autism, 139–42

babies, 28, 31–34, 53, 119, 123
  See also children
bachata, 236, 238
back pain, 104, 129–31
balance (sense of), 99, 104, 172, 175, 180–81, 187–88
Balboa (dance), 247

ball de bot (Majorca), 207
ballet, 197, 234
  responses to, 48–49, 211–13
  for the visually impaired, 172–73
  See also ballet dancers
ballet dancers, 57, 61–62, 91, 136–37, 178
ballroom dancing, 223, 236–37
  See also partner dancing; and specific dances
ballu sardu (Sardinia), 135
belly dancing, 131, 182, 198, 245–46
blindness, 171–74
blood pressure, 98–99, 101, 155
blood sugar, 101
blues, 247
body awareness, 111–13, 163, 181, 188, 191
body image, 134–37, 138

299

Bokwa, 228
bolero, 236
bones, 104. *See also* spine
boogie-woogie, 247
brain, 114–15, 154, 169, 183
   movement and, 111, 117,
     207–11, 218–19
   music and, 105–7, 111
   tango and, 55, 188
   See also learning;
     neuroplasticity
break dancing, 239–40, 247
   See also hip-hop
breathing, 45, 99, 102–3, 126,
   130, 251–52
burnout, 155

cancer, 127, 142–43
capoeira (Brazil), 61–62, 137,
   175, 192
cardiovascular system.
   See heart health
cha-cha, 106, 137–38, 236
Charleston (dance), 18, 247
children, 56–57, 115–16, 189–93
   See also babies; teenagers
choreography, 58–59, 214–15
circulation. *See* heart health
club dance, 249–50
cochlear implants, 175–76
cognition, 54, 119, 183
   See also brain; dementia
communication, 52, 54, 55,
   57–58, 76–79
community, 83, 92–95, 108,
   221–23, 225–26

contact improvisation, 248–49
coordination, 100
   See also synchronization
cortisol, 154, 155, 157, 158,
   159–60, 162
country dancing, 184
creativity, 113, 118, 119, 165,
   192, 210–11
crying, 168–69

dancing
   for the blind, 171–74, 176
   for the deaf, 174–76
   dressing for, 152–53, 233
   meaning in, 214–19
   online, 251–52
   as role-play, 76–79, 192–93
   for seniors, 178–79
   and social change, 221–25
   as social contact, 180, 201–6
   as therapy, 89, 126–27, 163–
     64, 166, 167–68
Dancing With the Stars, 69, 134,
   171, 172
deafness, 174–76
dementia, 182–85
   See also brain; cognition
depression, 142, 164–71
diabetes, 100
digestion, 101
disco dancing, 249–50
dopamine, 29, 81, 110, 150–51,
   187
   and mood, 111, 150

Index of Health Concerns and Dances

eating disorders, 143
elderly, 178–85
emotions, 44–50, 56–57, 120
   painful, 116, 167–69, 208
   trying out, 144–47, 168
   See also specific emotions
endorphins, 95, 107, 111, 155, 168, 201
endurance, 98, 125, 126, 180–81
exercise, 98–99, 127, 134–36, 155, 191
   See also fitness
exhaustion, 155

falls. *See* balance (sense of); body awareness
fasciae, 100
fatigue, 147–50, 151–52
fitness, 98–99, 100, 126–27, 184
   See also exercise
flamenco, 86–88
flexibility, 98, 100, 104, 180
   cognitive/neural, 118, 183, 190, 224, 251
   of the spine, 128, 132
folk dancing, 82–83, 84–89, 135, 243–44
   See also group dancing; and specific dances
formation dances, 90, 242–43, 246–47
foxtrot, 236
freestyle dance, 58–59, 68, 92

gastrointestinal system, 101, 102

Greek dances, 86, 89–90, 142
grief. *See* sadness
group dancing, 158–59
   synchrony in, 91–92, 93–95
   traditional, 82–83, 84–85, 245
   See also folk dancing; and specific dances

haka (Maori dance), 84–85
hambo (Sweden), 135
happiness, 49, 169
   See also endorphins
hearing impairment, 174–76
heartbeat, 123–24, 134–35
heart health, 98–99, 101, 124–27
hip-hop, 137–38, 224–25, 239–40, 247

immune system, 100, 101, 108, 109, 124, 127–28
improvisation, 248–49
incontinence, 181–82
Indian dance, 58
Indigenous dances, 21–22
   See also folk dancing; and specific dances
inflammation, 100, 102, 128
insulin, 154
intelligence, 118
   See also cognition
interoception (body awareness), 111–13, 163, 181, 188, 191

jazz dance, 18, 181, 242–43
jitterbug, 247
jive, 236, 247
joints, 100–101, 132–33

Latin dance, 100, 106,
    199–200, 236–37
    See also salsa; Zumba
laughter, 79–80, 228
    See also happiness
learning, 54–55, 115–16, 119–20,
    177–78, 190–91
    See also cognition
Lindy Hop, 18–19, 178–79, 247
    See also swing dance
line dancing, 184

mambo, 236
meditation, 158–59, 160, 166,
    173
memory, 119, 185, 225–26
    See also dementia
merengue, 201, 236, 238
milonga. *See* tango
mirroring, 56, 60–66, 79, 120,
    207–10
modern dance, 45–46, 181,
    214–15, 216–18, 241–42
mood, 102
    See also emotions
motor skills, 53–54, 115–16,
    120–21
    See also autism; Parkinson's
    disease
movement
    backwards, 102–5

and the brain, 111, 117,
    207–11, 218–19
and emotions, 144–47, 168
and learning, 54–55
sensory impairment and,
    173, 175
watching, 206–19
    See also mirroring
muscles. *See* strength
music
    and the brain, 105–7, 111
    and memory, 185, 225–26
    swaying to, 19, 24–25,
        252–53

neuroplasticity, 117–18, 184–85
ngoma (Tanzania), 158–59
noradrenaline, 44, 154

old age, 178–85
oxytocin, 110, 201

pain
    in back, 104, 129–31
    brain and, 80, 208
    emotional, 116, 167–69, 208
    in joints, 132–33
    perception of, 95, 100
    therapies for, 142, 159
palliative care, 143
panic attacks. *See* anxiety
Parkinson's disease, 186–89
partner dancing, 65–81, 108, 172,
    201–6
    as communication, 52,
        76–79

Index of Health Concerns and Dances

and relationships, 66, 70–72, 80–81
and sex, 71–74, 200–201
See also specific dances
paso doble, 236
Persian (Iranian) dances, 198, 250–51
pole dancing, 198
posture, 99–100, 101, 128–32, 245
prolactin, 169
PTSD. *See* trauma
puberty. *See* teenagers

qigong, 158, 160
quickstep, 236

respiration. *See* breathing
rock 'n' roll, 27, 71, 247
rueda Cubana, 90, 238, 245
rumba, 27, 236

sadness, 50, 165, 167–69
See also depression
salsa, 79, 181, 199–200, 231, 236, 238–39
samba, 236
schuhplattler (Germany), 86
self-consciousness, 35, 37–39
See also teenagers
seniors, 178–85
serotonin, 44, 111, 166, 168
sex, 35–36, 195–201
partner dancing and, 71–74, 200–201
See also testosterone

shag, 247
sirtaki (Greece), 86
spatial awareness, 188, 190
spine, 128–29, 132
square dancing, 90, 184, 246–47
stamina, 98, 125, 126, 180–81
step dance, 90, 247
street dance, 239–40
See also hip-hop; salsa
strength, 99, 100, 101, 180
stress, 102, 127–28, 153–60
See also anxiety
stretching, 97–98, 102, 104, 183
swing dance, 31, 79, 117–18, 231, 247–48
older people and, 178–79
as social activity, 64–65, 83–84
as therapy, 89, 169–70
and weight loss, 137–38
See also Lindy Hop
synchronization, 91–92, 93–95
syncopation, 28–30

tai chi, 158
TaKeTiNa, 159
tango, 74–75, 81, 144–45, 213, 225, 235–36
benefits of, 148–50, 157, 166, 187–88
and the brain, 55, 188
See also partner dancing
tarantella (Italy), 155–57
teenagers, 37–41, 69, 166–67, 227–28

303

testosterone, 44, 73, 75, 157, 201
touch, 67, 108–10, 173
trauma, 143–44, 160–62
   See also anxiety; stress

vagus nerve, 101–3
vision impairment, 171–74
voguing, 222, 239–40

waltz, 28, 67–68
   See also ballroom dancing
weight loss, 133–38

yoga, 102, 158

Zumba, 116–17, 135, 236–37